T0226094

Advanced Technologies and Societal Change

This series covers monographs, both authored and edited, conference proceedings and novel engineering literature related to technology enabled solutions in the area of Humanitarian and Philanthropic empowerment. The series includes sustainable humanitarian research outcomes, engineering innovations, material related to sustainable and lasting impact on health related challenges, technology enabled solutions to fight disasters, improve quality of life and underserved community solutions broadly. Impactful solutions fit to be scaled, research socially fit to be adopted and focused communities with rehabilitation related technological outcomes get a place in this series. The series also publishes proceedings from reputed engineering and technology conferences related to solar, water, electricity, green energy, social technological implications and agricultural solutions apart from humanitarian technology and human centric community based solutions.

Major areas of submission/contribution into this series include, but not limited to: Humanitarian solutions enabled by green technologies, medical technology, photonics technology, artificial intelligence and machine learning approaches, IOT based solutions, smart manufacturing solutions, smart industrial electronics, smart hospitals, robotics enabled engineering solutions, spectroscopy based solutions and sensor technology, smart villages, smart agriculture, any other technology fulfilling Humanitarian cause and low cost solutions to improve quality of life.

Ninni Singh · Vinit Kumar Gunjan ·
Jacek M. Zurada

Cognitive Tutor

Custom-Tailored Pedagogical Approach

 Springer

Ninni Singh
Department of Computer Science
and Engineering
CMR Institute of Technology
Hyderabad, Telangana, India

Vinit Kumar Gunjan
Department of Computer Science
and Engineering
CMR Institute of Technology
Hyderabad, Telangana, India

Jacek M. Zurada
Department of Electrical and Computer
Engineering
University of Louisville
Louisville, KY, USA

ISSN 2191-6853   ISSN 2191-6861   (electronic)
Advanced Technologies and Societal Change
ISBN 978-981-19-5199-2   ISBN 978-981-19-5197-8   (eBook)
https://doi.org/10.1007/978-981-19-5197-8

This Springer imprint is published by the registered company Springer Nature Singapore Pte Ltd.
The registered company address is: 152 Beach Road, #21-01/04 Gateway East, Singapore 189721,
Singapore

Preface

Artificial intelligence is an advanced field of research. It is particularly used in the field of education to increase the effectiveness of teaching and learning techniques. With the advancement of internet technology, it has been observed that there is a rapid growth in distance learning modality through the web. This mode of learning is better known as the e-learning system. These systems present low intelligence because they offer a preidentified learning frame to their learners. The advantage of these systems is to offer to learn anytime and anyplace without putting emphasis on a learner's needs, competency level, and previous knowledge. Every learner has different grasping levels, previous knowledge, and preferred mode of learning, and hence, the learning process of one individual may significantly vary from other individuals.

Ongoing research and development initiatives have led us to the origin of intelligent tutoring. It has gained immense popularity in current times specifically due to the advancement of intelligent tutoring systems (ITS). Research in psychology, education, and computer science (AI and machine learning) fueled the foundation of the field of intelligent tutoring systems. Thus, ITS aims to cognize the learner's needs, and grasping levels, and offer the learning material that best suits the learner's requirements. ITS acts as a cognitive tutor that not only solves the learner's issues (hints and feedbacks) but also keeps an eye on the learner's performance and activity during earning and deduces the competency level of learner's in the particular subject domain. It seeks to determine the learner's cognitive state of mind. Thus, identification of the cognitive state of learners makes a computer-assisted learning system an intelligent tutoring system. There is progressive growth of a computer-assisted learning system to various forms of web-based learning systems and a further advancement to form a personalized tutoring system.

This book illustrates the design, development, and evaluation of the personalized intelligent tutoring system that emulates the human cognitive intelligence by incorporating the artificial intelligence features. This book provides a complete reference for novice students, researchers, and industry practitioners interested in keeping abreast of recent advancements in this field. This book is suitable for postgraduate students, researchers, scholars, and developers willing to gain cognitive intelligence knowledge. This book encompasses cognitive intelligence and artificial intelligence which are very important for deriving a roadmap for future research on intelligent systems.

Hyderabad, India   Ninni Singh
Hyderabad, India   Vinit Kumar Gunjan
Louisville, USA   Jacek M. Zurada

Contents

Chapter 1
Introduction

1.1 Introduction

A successful learning and teaching process is made possible by using artificial intelligence (AI). Many computer artifacts have grown incrementally over the past few years due to extensive research in this domain. Intelligent tutoring system (ITS) [1] is one of many types of computer artifacts. There has been a convergence of three academic areas, namely computer science, psychology, and education (Fig. 1.1). Psychology is concerned with how students behave when working with a tutoring system. Cognitive intelligence and human tutoring abilities can be emulated by using digital technology covered by computer science. This includes the collection of knowledge and the method of delivering it to the intended audience.

Intelligent tutoring system (ITS) results from research in psychology, education, and computer science (AI and machine learning). Its goal is thus to understand and meet the learner's demands and provide the learning content that best meets the learner's expectations. ITS acts as a cognitive tutor that addresses the learner's concerns (hints and feedback) keep a watch on the learner's performance and activity while learning, and deduces the competency level of the learner in the given subject area. Its goal is to ascertain the mental state of the student. A computer-aided learning system becomes an intelligent tutoring system when it can its students' cognitive state. Personalized tutoring is becoming more and more common as computer-assisted learning evolves into a variety of web-based solutions.

When compared to other web-based learning systems, ITS stands out. An online learning system allows learners to follow a predetermined sequence of learning materials. Still, it does not provide tailored learning environments or the necessary personal supervision (hints and feedback) that learners need to succeed in their education.

In the literature, there are numerous ITSs/E-learning systems, each with a particular architecture [2–7].

Numerous instructional technologies in science, technology, engineering, and mathematics (STEM) have been implemented. Knowledge of well-documented

Fig. 1.1 Intersection of disciplines leading to the birth of an intelligent tutoring system

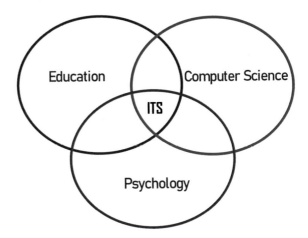

fields, such as electrodynamics, physics, SQL, C++, and physiology for medical students, has been found in the literature to be used in the development of the ITS [3, 8–13]. Until now, there has been no ITS where domain knowledge is based on domains that are not well-documented, i.e., not available in explicit form, i.e., experiential knowledge.

Learning by experience is a type of knowledge known as experiential knowledge. In more specific terms, it is referred to as "activity-based learning." As a result of your experiences, you will be able to learn and grow as a person (see Fig. 1.2). Activist learning, adventurous learning, cooperative learning, and active learning are examples of this type of learning.

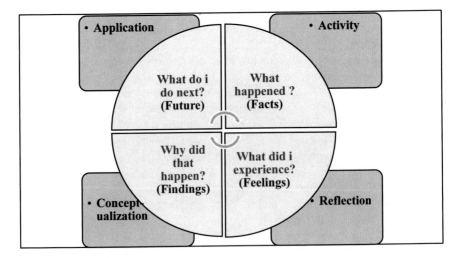

Fig. 1.2 Features of experiential learning

   Seismic data interpretation is a sort of experiential learning in this study's "Seismic Data Interpretation." The field of geophysics known as seismic data interpretation (SDI) is concerned with interpreting seismic data. Seismologists (specialist scientists) use the seismic SEG-Y Map to sketch out the earth's sub-surface structure during exploration. Seismologists utilize their expertise to deduce the presence of hydrocarbons from the sub-surface structure. Due to the difficulty of the interpretation, it takes longer to reach a judgment that is supported by sufficient evidence. It is difficult to persuade other seismologists because of a lack of evidence. As a result, the results drawn by seismologists are open to question. The lack of general guidelines is to blame for this ambiguity. Hands-on training is how seismologists hone their interpreting skills. Thus, this knowledge is based on personal experience and is highly personal. SDI is therefore considered to be a form of tacit knowledge.

   Technology that can help the adaption process in e-learning systems, such as curriculum sequencing approaches and cognizing the student's psychological state, has been researched and is still being done. A learning route is a system for organizing and presenting learning materials and activities to students to help them achieve their learning objectives. Using curriculum sequencing, a student's learning path can be tailored to their specific needs by automatically picking the best teaching format for each student. Depending on the learner's characteristics, this journey may look different for each person. This will allow each learner to have a unique learning experience, which will help them retain the information better. Every student has a prior understanding of the subject matter of the course. As a result of this work, it can be concluded that providing learning content based on the prior knowledge and preferences of the learner improves learning outcomes and student engagement.

   As part of the personalization, an adaptive intelligent tutoring system considers the student's feelings. For example, in Wolcott [14], it is suggested that students' psychological state is identified by non-verbal ways such as facial expressions, body language, and eye contact that signal the degree of knowledge and involvement. An intelligent tutoring system that incorporates emotion recognition will therefore be able to provide better recommendations, better tutoring, and make tutoring sessions more worthwhile.

1.2 Need of an Intelligent Tutoring System

According to a recent study, Internet technology has led to a dramatic increase in the use of web-based distance learning. The e-learning system is the more common name for this approach to education. There is a lack of intelligence in these systems because they provide learners with a predetermined learning framework. Learning can be done at any time and in any location thanks to these systems, which don't emphasize a learner's needs, competence, or prior knowledge. Learning styles, prior knowledge, and preferred learning methods vary widely among students, making it impossible to generalize across groups.

According to the literature, if a learner's past knowledge and preferences are known, then learning can be proposed in a manner that promotes the learner's learnability [15, 16]. Therefore, learning style, learning profile, and prior knowledge are all factors that contribute to the overall efficacy of an Individualized Instructional Plan, which is defined as one that is "custom-tailored" for each student. One of the new characteristics of ITS is the recommendation and alignment of learning material based on the learner's prior knowledge, competency level, and absorption level. One of the most significant issues in ITS is to include this cognitive intelligence into the system [17].

There are various challenges in knowledge-based ITS:

i.   Predeciding the rules for creating tutoring strategy
ii. Even in the best-case scenario, experts may have to come up with a number of regulations to handle the situation. Think of a scenario in which students must discover how many different ways there are to get from point A to point B; experts must then devise numerous rules (based on all conceivable scenarios) to solve the same problem.

The effort of building the knowledge base is a costly and time-consuming one. If the knowledge is very individualized, it takes more effort to create knowledge units (capsules). As a result, this knowledge can be found in a tacit state. Tacit knowledge elicited from experts is a serious bottleneck since experts conclude without appropriate reason. Consequently, obtaining and conveying experiential knowledge is a huge difficulty.

ITS has various research gaps, out of these few of them are addressed below. Following are the few gaps of ITS:

i.   Issues in the acquisition of tacit knowledge (seismic data interpretation) from the experts.
ii.   Challenges of explication of tacit knowledge.
iii. Non-availability of knowledge repository for tacit knowledge of seismic data interpretation domain.
iv.   Lack of generation of exclusive tutoring strategy (learner-centric).
v.   Lack of sequencing of learning material as per learner preference and previous knowledge.
vi.   Lack of adaptivity through pedagogical recommendation.

1.3   A Growing Field Intelligent Tutoring System

Computer-aided instruction (ICAI) and intelligent tutoring systems (ITS) give one-on-one education equivalent to that provided by an excellent human tutor. These are the primary goals of intelligent computer-aided instruction. The development and testing of models describing the cognitive processes involved in instruction is another goal of ITS research. However, despite these two objectives being intertwined, various research priorities may be placed on them.

The first ITS was built in the 1970s (for an overview of these early systems, see Wenger 1987).

The discipline of intelligent computer-aided instruction (ICAI) or intelligent tutoring system (ITS) was born as a result of further advancement [1]. In CAI systems, the learning material is stored, which may then be accessed by the learner in various ways (representation). There are various drawbacks to using these systems. Instead of emphasizing education's qualitative nature, these systems emphasized its quantitative nature (teacher-centricity). CAI systems use a rudimentary tutoring method, which results in less contact between the CAI tutor and the learner.

ITS aims to fix the flaws in the CAI system. ITS is also referred to as a cognitive tutor since it provides learning material tailored to the learner's specific learning style. These systems use artificial intelligence (AI) approaches to provide adaptive tutoring [18]. With the advent of ITS, tutoring sessions are made more efficient since it caters to students' individual needs, guide and monitor their development, and offer the feedback they require.

California Assisted Instruction (CAI) was the first intelligent tutoring system (ITS) to be upgraded [19], according to developed intelligence-enhanced computer-aided education systems that successfully communicate learning content to students through successful interaction with the system's artificial intelligence. So these systems only partially imitate human cognitive intelligence, in the sense that they only provide relevant help based on learner action when the system is activated. On the other hand, this technique does not adequately address all aspects of the learner model. Furthermore, the amount of information gained by learning grows too large to be immediately included in the code.

The ITS developed by Uhr [20] generates vocabulary and arithmetic problems; however, it cannot adapt to and model student requirements. Several adaptive systems developed by researchers [21–23] have made significant progress in this area [24].

The learner model and the pedagogy model, on the other hand, were not adequately studied. The learners' demographics, performances, and participation in action activities are not successfully stored. These were the initial examples of ITS. Another method, known as "drill and test," was also developed. Learners' performance and the reaction of learners serve as critical criteria in these systems' identification and recommendation of the next set of exams and the next group of upcoming learning content to be presented to the student. Despite this, these systems cannot provide the necessary guidance in the form of feedback, resulting in a research gap in developing a new era of the learning system.

The growth of learning systems has resulted in significant modifications to the design of the information technology services. Further improvements in the pedagogy model resulted in the tailoring of the learning content to the level of competency of the learner population. As a result, the emphasis on implementing artificial intelligence technologies has shifted from generalization to the fine-tuning pedagogical recommendation and learner feedback systems.

Physics is a subject that has received a great deal of attention [24]; have developed an intelligent tutoring system that is based on physics as its knowledge domain. They make decisions based on the results of a Bayesian network. In the Andes, the physics problem is dissected and utilized to form a Bayesian network of interacting elements. With the Andes, you get an intelligent function that assesses performance factors, predicts learner actions, finds applicable methods, and proposes them. Furthermore, this network contributes to determining the most efficient learning path available [26].

Fuzzy techniques determine the learner's prior knowledge, and data structure concepts (graphs) sequence the learning information in IntermediActor.

SQL tutor employs an artificial neural network (ANN) to determine which learning material to present next. Therefore, learning psychology and learner responses to a task are taken into consideration throughout learning sessions, as well as during assessments Mitrovic [23].

Mooney designed C++ Tutor as an ITS. This tutoring system provided questions in feature vectors, which the student then answered. The learner's responsibility is to identify and label the vectors in this situation. As part of the rule base development, they use the NEITHER algorithm, which facilitates the deduction of learner perceptions from solutions provided by the learner. The term "THEORY REVISION" refers to the entire process of revising one's theory. After this procedure, the system displays the learners' misconceptions in question.

Evens developed CIRCSIM Tutor, an ITS that is dialogue-based. The domain model represents the subject domain of physiology. It is separated into four components: tutoring history, learner history (reaction), performance, and learner solution [27]. Following an evaluation of existing ITS, the authors [28] found that the following research gaps needed to be addressed:

- The absence of a custom-tailored course coverage plan
- Creating a customized knowledge pool
- Customizing learning content and learning materials
- The customization of the learner.

VisMod is an ITS developed by Freedman et al. [29]. According to the developers, this system has a three-level hierarchical architecture and provides the user with learning content across a wide range of topic fields.

The most notable feature of ITS that distinguishes them from other artificial intelligence applications such as expert systems (see Expert Systems in Cognitive Science) is the assessment of the current state of the student's knowledge (see Assessment of the Present State of Knowledge in Cognitive Science).

1.4 Effectiveness of ITS

The best way to help students struggling with ideas or exercises is to hire a one-on-one tutor. Research shows that one-on-one tuition with a genuine person is effective [30–32]. Students can benefit from human instructors in a variety of ways. While allowing students to complete as much work as possible, good human tutors also keep them on track to find solutions [33]. Studying independently can also help students develop their general knowledge and critical-thinking abilities. But it could take a long time and a lot of effort. A one-on-one tutor can demonstrate to a student how to employ an approach that works and what doesn't work to overcome obstacles. It is common for tutors to challenge and excite pupils while also allowing them to feel like they are in command. Instead of only telling pupils what to accomplish, human tutors provide recommendations and inspiration. This inspires kids to persevere in the face of adversity. Students benefit greatly from the one-on-one interaction with real tutors because they receive immediate feedback as they work through challenges. An ITS must interact with students in the same manner as a human tutor to provide feedback comparable to that of a human tutor. This raises how an ITS can assist students in the same way a human instructor can.

As students work through a challenge, every step they take must be documented for the model. The system can detect when a student has strayed from the path and intervene to help the learner get back on track by keeping track of each step individually. Students can get help and comments when they make mistakes and hints when they don't know what to do next. Comparing the measures used by a student to those of a rule-based domain expert can help determine how they solve a problem. The system keeps track of a student's progress in the model tracing method. The system's job is to help students understand what went wrong when they make mistakes or incorrect assumptions. It does not sound as if the system monitors the student's progress. Studies suggest that students benefit from tracing their models, such as the visual exposition of geometric concepts in geometry tutor and the explicit instruction of LISP (GIL) [33]. As it turns out, pupils benefit from using model tracing tutoring methods [33, 34].

It's critical to know whether or not ITSs deliver on their promise of improving student learning outcomes. Meta-analyses have examined ITSs to see how effective they are. This kind of research has been done recently, and the results are presented here to demonstrate what has been discovered. To see how effectively computer tutoring, human tutoring, and no tutoring worked, VanLehn conducted a meta-analysis in 2011 [34]. In this study, computer tutors were categorized according to the level of depth in their user interface interactions: answer-based, step-based, and sub-step-based groups. From 1975 to 2010, consider a wide range of research. Ten comparisons were drawn from 28 evaluation studies. According to the study, human tutors had an impact size of 0.79 compared to those who did not have a tutor. This is not as good as Bloom's 2.0 impact size [31]. It was also observed that step-based tutoring (0.76) was practically equal to human tutoring, whereas sub-step-based tutoring (0.48) was marginally better than no tutoring. In light of VanLehn's findings, researchers in tutoring should focus on making computer tutoring as effective as human tutoring, which has a 2.0 times greater effect than no tutoring. An investigation by Steenbergen and Cooper in 2013 examined the effectiveness of ITSs in helping children in grades K–12 acquire mathematics [35]. This study analyzed 26 studies that evaluated the effectiveness of ITSs and traditional classroom training. They observed that using ITSs for a short period did not affect how effectively students studied. At least one school year of use proved to improve the ITS performance. Students who did well in school looked less affected than those who did not.

Ma et al. [36] carried out a meta-analysis in 2014, they compared students' learning outcomes who used ITSs to those of students who did not. This study's purpose was to examine the effect sizes of various ITSs by considering factors like the type of ITS, the type of instruction (individual or small group instruction, for example), and the subject area (mathematics or chemistry, for example). Ma and colleagues examined more than 100 papers to determine the size and scope of this effect. The ITS environment was 0.42 compared to teacher-led and effective group teaching, 0.57 for computer-based instruction without ITS, and 0.35 for textbooks or workbooks. Individualized human tutoring (-0.11) and small group instruction (-0.11), on the other hand, did not affect how effectively participants learned (0.05). However, except for human coaching in small groups, Ma et al. discovered that ITSs were the best at teaching. Other factors such as student characteristics, domain expertise, and others contributed to differing ITS impact results. Researchers Kulik and Fletcher conducted an extensive review of 50 papers in 2015 to examine how ITSs and traditional classrooms effectively assist students in learning [37]. Study after study found that students who worked with ITSs performed better than those taught in the usual fashion. A moderate-to-high effect size of 0.66 was found in 39 of the 50 trials. However, on standardized examinations, the effect size was only 0.13. Researchers must discover solutions to questions because no one can agree on how well ITSs work. How effective are ITSs in the real world? What factors have the most bearing on student progress in ITSs? So, what can be done to improve the ITSs?

1.5   Intelligent Tutoring System Architecture

What is the importance of architecture in today's world? When it comes to ITS deployment, traditional trinity models, according to Self, "will create barriers and will limit the breadth of ITS studies" in 1990. The classical trinity does not meet the criteria for a sound philosophical basis when used as a starting point for creating ITS theory development [28]. This is a very crucial issue to take into consideration. A "trinity" or "four-component architecture" is a type of architecture that is frequently referred to as a "trinity" or "trinity architecture" (Fig. 1.1). It is critical to understand the symbolic significance of the building. Therefore, even though the system has been broken down into four distinct components, the integration challenge has remained unsolved. The use of one component over another in the context of computation and control is a common occurrence in systems. Is it safe to say that the information included inside each element may be relied upon?

It's critical to know whether or not ITSs deliver on their promise of improving student learning outcomes. Meta-analyses have examined ITSs to see how effective they are. This type of research has been done recently, and the results are presented here to demonstrate what has been discovered. To see how effectively computer tutoring, human tutoring, and no tutoring worked, VanLehn conducted a meta-analysis in 2011 [34]. In this study, computer tutors were categorized according to the level of depth in their user interface interactions: answer-based, step-based, and sub-step-based groups, from 1975 to 2010, considered a wide range of research. Ten comparisons were drawn from 28 evaluation studies. According to the study, human tutors had an impact size of 0.79 compared to those who did not have a tutor. This is not as good as Bloom's 2.0 impact size [31]. It was also observed that step-based tutoring (0.76) was practically equal to human tutoring, whereas sub-step-based tutoring (0.48) was marginally better than no tutoring. In light of VanLehn's findings, researchers in tutoring should focus on making computer tutoring as effective as human tutoring, which has a 2.0 times greater effect than no tutoring. An investigation by Steenbergen and Cooper in 2013 examined the effectiveness of ITSs in helping children in grades K–12 acquire mathematics [35]. This study analyzed 26 studies that evaluated the effectiveness of ITSs and traditional classroom training. They observed that using ITSs for a short period did not affect how effectively students studied. At least one school year of use proved to improve the ITS performance. Students who did well in school looked less affected than those who did not. Ma et al. [36] carried out a meta-analysis in 2014, they compared students' learning outcomes who used ITSs to those of students who did not. Ma et al. discovered that ITSs were the best at teaching. Study after study found that students who worked with ITSs performed better than those taught in the usual fashion. Other factors such as student characteristics, domain expertise, and others contributed to differing ITS impact results. Researchers Kulik and Fletcher conducted an extensive review of 50 papers in 2015 to examine how ITSs and traditional classrooms effectively assist students in learning [37]. A moderate-to-high effect size of 0.66 was found in 39 of the 50 trials. However, on standardized examinations, the effect size was only 0.13. Researchers must discover solutions to questions because no one can agree on how well ITSs work. How effective are ITSs in the real world? What factors have the most bearing on student progress in ITSs? So, what can be done to improve the ITSs?

The conventional ITS design consists of five key components: the learner model, pedagogy model, expert model, knowledge/domain model, and learner interface model (see Fig. 1.3).

Fig. 1.3 Architecture of typical ITS

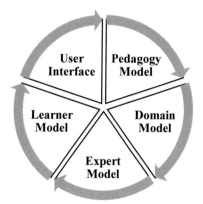

1.5.1 Learner Model

Freedman et al. [29] state that the learner model captures information about the learner, such as prior knowledge and learning styles and mistakes and misunderstandings. Learner models are the foundation of an IT infrastructure. Students' cognitive and emotional states and how they change over time should be included in a student's progress report. Throughout the training, data about student behavior and performance is also stored in the system (such as correct responses, the hint opted, and incorrect answers). Most commonly, the student model is thought of as a dynamic model that serves multiple purposes. Wenger [2] gave the student model three primary functions: It is possible to analyze the learner's knowledge and pedagogical strategies for imparting following domain information by acquiring explicit and implicit (inferred) facts about the learner. These data, in turn, must be used to create a representation of the student's knowledge and the student's learning process. A student model can play six key roles, according to [28].

Many of these positions and responsibilities have evolved since then, both in terms of their breadth and their depth. In addition to the tactical decisions made in 1 and 2, the tutorial strategy will need to undergo major adjustments to address students "incomplete" knowledge, which can be addressed in three ways: (1) correctively, (2) elaboratively, and (3) strategically. Depending on the purpose, it can be diagnostic, predictive, or evaluative. Diagnostics assist in identifying problems with a learner's knowledge, while predictive analytics help predicts how a student will respond to tutorial actions.

According to [1], there are six ways to employ the student model. Correlative instruction is aimed at helping students identify and remedy errors in their prior knowledge. The second type is more extensive because it fills in what the learner doesn't know about the subject matter. The lesson plan can be adjusted to reflect the learner's progress as they progress. Fourth, diagnostic, because it aids in the student's ability to detect errors in their knowledge. Predictive in nature, the fifth kind aids the educator in determining how the student will respond to changes in the system. An evaluative assessment gives teachers a sense of how well their students are performing. To gain insight into the pupil, the model can be applied. The system should be able to learn about the student's behavior based on this model, even if it cannot be seen. It should figure out what the student doesn't know by looking at what they do. Domain knowledge is likely to influence the presentation of the student model. A student model can incorporate assessments of how effectively the knowledge is learned by breaking it down into its component elements. Students' learning can be compared to that of experts using this method. As a result, a student's education can be tailored to their specific needs. Keep in mind that bad behavior isn't always the result of a lack of knowledge. The tutoring method becomes more difficult to use when the topics need to be taught vary over time. Because remediation may be necessary, a student model must depict the student's alleged incorrect knowledge. The student model must be tested. Predicting how a pupil will respond is made more accessible by this method. In the end, this is what allows this critical ITS architectural component to work well with students. Correcting misunderstandings, providing tailored feedback, and offering suggestions for how to develop new skills are all examples of these kinds of interactions [8, 9]. Creating a student model is not an easy task. People's answers to specific questions should play a role in the outcome. Is there anything you can tell me about the student? How can a learner tell what to do next when faced with difficulty? To create a student model, you should ask yourself these questions. To begin with, determine the extent to which the learner is familiar with the components of the mechanism. The second step is to assess the learner to determine how well a student comprehends the mechanism. Ultimately, it is crucial to decide what instructional approaches a student used when they came up with a solution. There are many considerations when designing the student model [38]. There are numerous characteristics of students to consider. To create an accurate student model, the system must consider both the student's fixed characteristics and their ever-changing characteristics. Email, age, and mother tongue are examples of static factors established before the learning process begins. Dynamic features are determined by the way students interact with the system. Each student's unique dynamic traits must be identified to tailor the system to their needs, according to [39]. Knowledge and abilities, mistakes and misunderstandings, learning styles and preferences, emotional and cognitive aspects, and meta-cognitive factors are some of the dynamic features that are constantly changing. This is an example of the terms "knowledge" and "learning style" or "preferences" being used interchangeably to describe how a student chooses to acquire new information (e.g., graphical representation, audio materials, and text representation). Students' emotional states, such as anger, happiness, sadness, or frustration, are all examples of affective elements. Students' cognitive abilities include focusing, learning, solving issues, and making decisions. People's willingness and ability to ask for help, regulate their behavior and assess their skills are all meta-cognitive traits [39, 40]. There were a lot of steps to creating the student model. These methods have been tried before and will be discussed in detail in the following sections.

1.5.1.1 Overlay Model

Stansfield, Carr, and Goldstein invented the overlay model in 1976. One of the most popular student models, it has been adopted by numerous tutoring systems. Assumes that what kids know is a small portion of what professionals know. A student's knowledge is considered lacking if they behave in a way that differs from those in the domain. For that, it's all about bringing them closer together [39, 41]. As a result, the overlay model illustrates how well the domain's elements are known. Boolean values indicate whether or not a pupil understands an aspect. The current overlay approach uses a qualitative measure to illustrate how much knowledge a student possesses (good, average, or poor). Using this strategy allows pupils to learn precisely what they need to succeed. On the other hand, this methodology has difficulty in the learner coming up with an alternative solution to a problem. In addition, the student may have "misconceptions" about the subject matter that aren't recorded in the student's domain knowledge. Carmon and Conejo have provided learner models for their MEDEA system, an open ITS framework [42]. MEDEA employed a conventional overlay paradigm to teach pupils what they learned and how it affected them. The learner model consisted of two parts: a model of knowledge and an attitude of the learner. Personal and technical characteristics of pupils, their preferences, and so on are included in the attitude model. These questions were posed to the learner before starting the learning process. The learner knowledge model was used to determine the student's knowledge and performance. These elements were altered during the learning process. Every topic in the domain is assigned a score based on how well the learner thinks they understand it [42]. People and computers alike may quickly scan a system's domain ontology thanks to InfoMap. A buggy model and an overlay model are used to determine what information is missing [43]. ICICLE is another ITS that uses the student model with the overlay model (Interactive Computer Identification and Correction of Language Errors). The purpose of the system is to employ natural language processing to aid students in learning the grammar of written English. A student overlay model is used by ICICLE to determine how well a user understands various aspects of grammar. Using this information can forecast which grammar rules the user is most likely to follow while creating language. ITS for computer programming called DeLC (Distributed eLearning Center) uses Kumar's overlay student model in 2006 [44] for distant and electronic education. The "overlay student model" was employed to determine the user's level of knowledge. There was a second way to model how learners used training resources, their routines, and their behavior during the learning process: the stereotype approach [45]. The LS-Plan is a framework for personalizing and adapting e-learning systems. Bloom's Taxonomy is used, and a qualitative overlay model is employed. A further tool used by LS-Plan is a bug model [46], which identifies issues users have with the software. PDinamet uses an overlay model to store concepts that the student has already learned or has yet to acquire when teaching physics in high school. Consequently, a tutor can consider a student's skill level and previous learning activities when suggesting a topic to them [47].

The overlay model ignores the students' incorrect information and cognitive needs, learning preferences, and learning styles. As a result, the overlay model is often used with other strategies, such as stereotypes and fuzzy logic.

1.5.1.2 Stereotype Model

Another method through which kids are frequently utilized as models is through stereotypes. Virvou et al. [48] developed the first system for modeling students based on stereotypes, the GRUNDY system. As Virvou et al. explains, "When we talk about stereotypes, we're referring to a collection of characteristics frequently seen in the same group of people. A smaller number of observations can yield much more accurate conclusions than a more significant number. However, these findings must be treated as defaults that can be altered by individual observations [49, 50]. Stereotypes are founded on the concept that all prospective consumers can be grouped based on a few common characteristics. A stereotype is a group that falls into this category. Users who join the site will be categorized according to their traits, as long as those traits fit within the stereotype. There is a wide range of options for students regarding learning from course materials in most information technology systems (ITSs). As a result, students may choose to study material that is too difficult or too easy for them, or they may skip over parts of the course.

Additionally, the ITSs should consider what students already know when creating, selecting, and assembling materials. To get a student model up and running, it is critical to use stereotypes to group students together. In other words, stereotypes help students learn, and teachers teach by reducing cognitive overload. The system may ask a few questions to get started in the student model [39]. As an example, consider a system that teaches Python programming. To determine which stereotypes children fit into, the system may begin interactions with them by asking them questions. You might also inquire about the student's C++ programming expertise. The system will presume that a student who is an expert in C++ is familiar with basic programming concepts like loops, while loops, and nested loops. As a result, the system will assign this student to a group of peers who share their familiarity with the fundamental concepts of computer programming [49]. Many adaptive tutoring systems use stereotypes to model students, and they often incorporate additional methods [49]. Inspire is a technology-enhanced instructional system (ITS) designed to cater to the needs of each student.

There are four ways to categorize someone's knowledge: insufficient, marginally adequate, or adequate. The fuzzy logic method is utilized to help pupils diagnose and deal with stereotypes [50] through the Web Passive Voice Tutor, another ITS that employs stereotypes to teach individuals how to use the passive voice in English. Machine learning and stereotypes were employed to ensure that each student received the appropriate education and feedback. The model is set up for a new student using a fresh combination of stereotypes and a distance-weighted k-nearest neighbor method [51]. The goal of AUTO-COLLEAGUE, a computer-supported collaborative learning system, is to assist users in learning UML in a way that is both individualized and adaptive. After discussing stereotypes and disturbances next, AUTO-COLLEAGUE employs a hybrid student model.

Three aspects of the user are stereotyped: (the level of expertise, the performance type, and the personality). Another ITS is CLT, which uses stereotypes to create student profiles. CLT teaches C++ iterative structures (while, do-while, and for loops). CLT stereotypes are based on verbal, numerical, and spatial abilities, each rated as low, medium, or high. When using the stereotype technique, Chrysafiadi et al. suggest that it is feasible to get a lot of information about a given user without interviewing them individually [39].

Additionally, it is simple and easy to keep track of stereotype-related information. On the other hand, stereotypes are problematic since they are based on the subjective opinions of a small number of people, usually other users or experts. Stereotypes are often incorrect in their depiction of the individuals they refer to. Numerous studies show that stereotypes aren't always accurate. Stereotypes have two extra drawbacks. Before they engage with the system, users must first organize into groups. Because of this, certain groups may no longer exist. The designer still has to create the stereotype, which takes time and is susceptible to errors, even if there is a class available.

1.5.1.3   Perturbation Model

The perturbation student model has supplemented the overlay model. To help students remedy their mistakes, the overlay approach displays their information as a subset of the expert's knowledge, which is easier for them to see and understand [52]. The perturbation model includes misconceptions or a lack of understanding, referred to as "poor knowledge" or "wrong beliefs." According to Martins et al. [52], replacing correct rules with incorrect ones will provide the perturbation model. The use of these words influenced the student's answers. Before and after the contact with technical information, there are various incorrect rules in the student's knowledge. The system gives a learner discriminating challenges to discover the user's incorrect rules. The "bug library" is where most pupils' blunders are kept. There are two ways to create a bug library: either gather the errors students make when utilizing the system (listing) or create an inventory of the most common errors students make when using the system (enumeration) (generative technique). This concept is superior to the overlay model in explaining why pupils behave the way they do. However, it is expensive to create and maintain [39]. Many adaptive tutoring systems use the perturbation approach to stimulate their students. The "perturbation model," or "buggy model," was utilized by Hung and his colleagues in 2005. Addition and subtraction faults are included in this model to determine what went wrong and why [43]. Precise, timely, and personalized feedback and assistance can be provided to students who are far away via LeCo-perturbation EAD's model [53]. In addition to stereotypes and the overlay model, Surjano and Maltby used the perturbation model to help students better repair their mistakes [54]. To help students better understand where they were making spelling mistakes, employed a perturbation student model in 2010.

1.5.1.4   Constraint Based Model

A constrained-based model (CBM) is a short-term model students use to figure out the present answer to a problem. The use of restrictions in CBM demonstrates both domain knowledge and student understanding [39]. The student's response can be determined by matching the relevant criteria of all constraints to the student's solution. All of the requirements for relevance have been met as well. Each step the learner makes is checked by the system, which identifies what went wrong and informs the student accordingly. If the student's answer is incorrect, the feedback tells them so, and then it identifies the specific domain principle that was violated [55]. It doesn't require a runnable expert module, which makes it easy to compute, and it doesn't require a lot of study on students' faults or intricate reasoning about where problems might have come from [56]. Tutoring systems in various domains have taken advantage of the CBM technique because of these advantages.

It is a web-based ITS for the SQL database language. Analyzes and provides constructive comments to students based on what they know and how quickly they can learn [57]. The ITS J-LATTE uses the CBM approach to teach its students a tiny portion of the Java programming language. For each solution that students submit, a list of relevant, satisfied, or maybe violated requirements is generated by the student modeler. The University of Hamburg's INCOM system assists students in taking a logic programming course. A weighted constraint is utilized to evaluate students' solutions [58] appropriately. Using the CBM student model, EER-Tutor teaches database topics and shows the student's level of understanding [55].

1.5.1.5 Cognitive Theories

Many academics have shown that utilizing cognitive theories to model students and determine where they went wrong makes tutoring systems more effective. A cognitive theory explains how people learn by analyzing how they think and understand. The Human Plausible Reasoning (HPR) Theory [56] and the Multiple Attribute Decision Making (MADM) Theory [57] are two cognitive theories that have been used in student modeling. Several plausible human deductions may be grouped into a collection of common patterns and techniques to change these patterns, known as Human Plausible Reasoning (HPR). Furthermore, it is a theory that may be used in any field. It was initially based on a database of answers to common inquiries. When students utilize HPR, they use RESCUER, an intelligent assistance system for Unix users, to emulate. To explain how a user concludes that the command they wrote is acceptable for UNIX, statements are transformed using the set of HPR transformations [58]. In addition to HPR, F-SMILE creates a model of the pupil. File-store manipulation intelligent learning environment is the acronym for F-SMILE. Its purpose is to teach newcomers how to utilize file-storage-related programs. Learners' cognitive states and personality traits are considered while creating an F-SMILE model of a pupil. The LM Agent uses a new combination of HPR and stereotypes in F-SMILE to make default assumptions about students until it has enough information about each one individually [59]. Another theory of how people think that has been used to build student models is Multiple Attribute Decision Making (MADM) [60]. MADM analyzes, ranks, or chooses among various options that have various, often contradictory, qualities to choose from. An intelligent learning environment for adults who are unfamiliar with the Windows 98/NT Explorer's GUI is provided by Web-IT. It's all done online. Personalized tutoring can be achieved using Web-MADM IT's and an age stereotype. Using a multicriteria decision theory, Alepis and Kabassi have developed a novel mobile educational system that improves the system's ability to recognize emotions. Bi-modal emotion recognition is used in the system, which employs the microphone and keyboard to interact with a mobile device.

1.5.1.6 Bayesian Network

Bayesian networks [39] are another well-known and proven method for demonstrating and justifying uncertainty in student models. Bayesian networks (BN) are directed, acyclic graphs containing random variables as nodes. A set of possible relationships between the variables is illustrated as arcs. The BN considers the context it models by analyzing action sequences, observations, and results [61]. Concerning the student model, nodes in a BN can display information like students' knowledge, misconceptions, emotions, learning styles, motivation, and objectives. BNs have been proven robust and beneficial in modeling situations involving knowledge. Theorists and system designers are interested in Bayesian networks because they are founded on sound mathematics and because they offer a natural method of displaying uncertainty using probabilities. As a result, BNs have been used in a range of areas, including medical diagnosis, retrieval of information, bioinformatics, and advertising, for various purposes such as debugging, diagnosis, forecasting, and classification [61]. People who want to use Bayes' theorem can do so with tools like GeNIe [62] and SMILE [63], making them simple to build and highly functional. The Andes is a type of ITS that helps with Newtonian physics [64, 65]. Andes' student model uses Bayesian networks to assess long-term knowledge, recognize plans, and estimate how students approach problem-solving. BN is also used in Adaptive Coach Exploration (ACE), an artificial learning environment for studying mathematical functions. Some other student designs make use of BN. The student model can provide tailored feedback on how a student investigates and determines whether a learner has problems exploring [66]. A Bayesian student model was also developed in an English grammar Assessment-Based Learning Environment. Pedagogical agents employ a Bayesian student model to provide adaptive feedback and arrange the sequence of activities [67]. E-teacher also uses a Bayesian student model to provide individualized assistance to e-learning students and automatically determine how each student learns [68]. By combining information on the causes and effects of emotional behavior [69], Conati and Maclaren used a dynamic Bayesian network to determine a great deal of uncertainty regarding the emotions of various users.

Similarly, a dynamic Bayesian network was integrated into PlayPhysics to determine the learner's emotional state based on their cognitive and motivational characteristics and behavior [70]. Technology Enhanced Learning Environment for Orthopedic Surgery (TELEOS) utilized the Bayesian student model to determine what the students knew and how they reasoned [71]. Crystal Island, a game-based learning environment in microbiology, also employed a Bayesian student model to forecast how students would feel by modeling how they now felt [72].

1.5.2 Pedagogy Model

An ITS gives each student personalized feedback based on the features in their student model. The teacher model, also known as the pedagogical module, is the most significant component of the system [73]. This model is similar to a human tutor in various aspects, including the capacity to choose what and how to educate. The duty of the tutor model is not only to guide the student like a tutor but also to make the learner's engagement with the ITS feel natural and seamless [74]. The pedagogical module should be able to answer questions like whether the next step for a learner should be a concept, a lesson, or a test. Other factors to consider are how to teach the student the topic, evaluate the student's work, and provide feedback to the student [75, 76]. In reality, the educational module connects with all other components of the expert and student models, acting as a bridge [77]. When a student makes an error, the educational module must provide feedback that describes the type of error, re-explains how to apply the rule, and aids the student as needed [78].

Built into every ITS is a pedagogical module that regulates student interactions. Furthermore, the instructor must choose what to teach the learner next, such as a topic or a problem to solve. The pedagogical model must consult with the student model to identify what the learner should concentrate on. This model makes these judgments based on the information contained in the student model and the information stored in the expert model regarding what the student has learned [58]. The educational module is responsible for the interaction between the learner and the system if the student needs assistance or makes an error that must be corrected. It accomplishes this by giving a series of feedback messages (akin to hints) or recommending that the student study a specific topic to boost learning performance. This section will learn about various instructional strategies that ITSs have utilized to provide content and implement necessary adjustments. Based on student activities, the learner model and domain model provide information for pedagogy model to make a strategic decision. In this model, the learner's preferences (profile and learning style) are used to identify the appropriate educational style based on the learner's preferences (the learner model). With input from domain and student models, tutoring decisions are made by the tutoring model. Decisions for intervention must be made based on a sound understanding of the facts. The tutoring model's functions also include content and delivery preparation. In an ideal world, tutoring decisions would be communicated to students in various ways, such as through Socratic dialogues, suggestions, and other types of feedback from the system. Snippets of knowledge are shared with the learner using the expert model. This model takes advantage of the expert's knowledge and abilities and portrays domain information in a way that aids in problem-solving skills development.

1.5.2.1 Decision Making in Cognitive Tutor and Constraint-Based Systems

Model tracing tutors (MTT) (also known as cognitive tutors) provide pupils with flag feedback, bug notifications, and a series of hints. The color of the feedback flag indicates to the learner whether their response was correct or incorrect (e.g., green for right and red for wrong). A "bug message" is linked to a student's incorrect answer to inform him or her of the type of error made. If the learner needs assistance, they can request a hint to receive the first in a series of ideas designed to provoke thought. When all leads have been delivered, the system will advise the pupil precisely what to enter [79]. ITSs that teach conceptual database design, such as KERMIT and SQLtutor, present the student with six levels of feedback: correct, error flag, hint, thorough hint, all errors, and solution. The right level indicates whether or not the student's response was correct. The error flag identifies the type of construct that includes the problem (such as an entity or a relationship). When a constraint is violated, a hint or a thorough hint is provided as feedback. The entire response is displayed at the solution level [73, 80].

1.5.3 Domain Model

The domain model is where ITS stores the knowledge or information it wants to use. To learn a new domain, one must first build a model of that domain (also known as "expert knowledge").

1.5.3.1 Cognitive Model

ITSs domain knowledge is traditionally modeled using the cognitive model. It has been utilized successfully in various ITS systems [81]. The domain's tasks are created by cognitive tutors, who apply a cognitive paradigm to their work. Many studies have indicated that cognitive tutors help students learn and improve their learning ability [82]. Math, physics, and computer programming are just a few of the fields where they've been employed [83]. To give pupils immediate feedback, cognitive tutors use a model of how the mind functions. For example, problem-solving ideas and methodologies can be described straightforwardly and precisely using this style. A rule-based paradigm provides pupils with a step-by-step approach to solving challenges. It can keep track of numerous methods to arrive at the correct answers and provide pupils with feedback on how accurate or wrong each step is (called "strategies"). Figure 1.4 [79, 84] displays both the correct responses and the common blunders made by pupils (Bug Libraries). Cognitive tutors have been developed following the ACT-R theory of how individuals think and learn [85]. ACT-R is predicated on the notion that knowledge can be either explicit or implicit. Many people believe that procedural

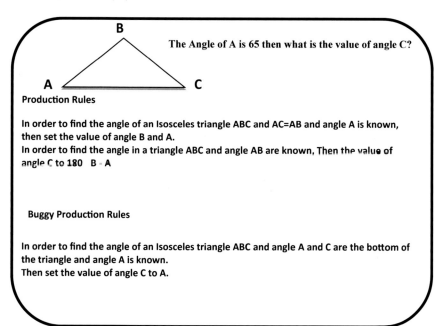

Fig. 1.4 Rules (Buggy) for determining size of angle

and declarative knowledge are distinct. For example, facts and ideas are connected in a semantic net or similar network to generate declarative knowledge.

On the other hand, procedural memory refers to our ability to recall specific steps in a process. As for if–then production rules, they are kept in a database. So, chunks and productions are the essential components of an ACT-R model [86, 87]. Model tracing, an algorithm, has been used to aid teaching through cognitive models. By comparing the student's answer to the model's, a tutor can determine whether or not it's correct. Students are right if their actions mirror those of the model. There must be an error if this isn't correct. An error is hypothesized when a student steps.

It does not match any rule or one or more of the buggy rules. A student's ability to follow a production rule might be viewed as a skill. As a result, over time, the model will be able to determine which abilities a pupil has already learned (knowledge tracing). In other words, students can use knowledge tracing to see what they have learned by solving a problem [87, 88]. Use the knowledge tracing approach, also known as cognitive mastery learning [89], to determine how likely a pupil is to know each skill. Each parameter associated with a skill is considered in the model, which calculates a student's probability of learning it. Cognitive mastery learning has long been recognized as having a significant impact on students' learning abilities. Using this model, data mining techniques like LFA and PFA have been employed to make ITSs even better [90, 91]. Cognitive tutors have had a big impact on students' education, but model tracing tutors haven't had the same impact in schools or other places like corporate training.

Software components such as an interface, a curriculum, a system for managing interactions between learners, and a package for teachers to utilize to report progress are required to develop cognitive tutors that are full and effective. As a result, the process involves a team of experts, which increases the cost and time [92]. There isn't much to do for these tutors with these two needs. Model tracing tutors can now be created with authoring tools that share the same characteristics, saving time and money. An example of this is Cognitive Tutor Authoring Tools (CTAT). Full ITSs have been created using CTAT, and no code has been written. There is a new category in the IT industry called example training Tutors [93].

1.5.3.2   Expert Approach

Adding an expert system to an ITS provides a third way to represent and apply domain knowledge for decision making. This information technology method can benefit from rule-based expert systems, neural networks, decision trees, and case-based reasoning can benefit from this information technology method. An expert system can solve a problem since it can make decisions and model skills like an expert. [94] Constrained and cognitive models can only deal with restricted domains, but expert systems can represent and reason about a much wider one [95, 95] demonstrated that an expert system strategy should comprise two modes. The expert system must first generate expert solutions. Comparing these solutions to the solutions that the learner came up with should then be possible. We can look at GUIDON [96] as an example of an ITS that employs this paradigm. Expert system technology can be employed in ITSs to compare the learners' solutions to the ideal answers. Systems like AutoTutor [97] and DesignFirst-ITS [25] can meet the second mode. There are some drawbacks to the expert system technique, as [78] points out: "This is especially true in domains where there aren't clear definitions of what constitutes an expert system. In some cases, expert systems cannot explain their judgments or provide suitable learning opportunities."

1.5.4   Expert Model

Expert knowledge is represented in several ways. One means of conveying information is through the use of facts and procedures. Among many functions are a repository of specialized knowledge, a yardstick by which students' work is measured, and a tool for spotting mistakes. As a result, it may be arranged into an overall curriculum, which includes all knowledge pieces connected according to the pedagogical sequence. The curriculum can be structured in various ways, including hierarchies, semantic networks, frames, ontologies, and production rules, depending on the level of information required for each knowledge unit.

Important issues include the ability to reason using a model and the gradual adaptation of the reasoning explanation to the learner's specific needs. An interface component is a tool used to facilitate communication between a student and a tutor in a classroom setting. In addition to simulations, hypermedia, microworlds, and other forms of learning environments, this component provides access to the domain knowledge elements.

1.5.5 Learner Interface

Using a learner interface paradigm, a learner can communicate with the learning system and interact with it. This approach incorporates both interactive media and graphics to reach out to students.

1.6 Book Contributions

An ITS's design, development, implementation, and evaluation were the emphasis of this book. As a result, the ITS covered in this book is dubbed "Seismic Data Interpretation" because of the domain knowledge. Due to the lack of thumb rules, this area is extremely experienced or tacit and relies mainly on seismologists' interpretation powers, capacities, and experience. It takes a long time to accumulate all of this knowledge. As a result, new seismologists must either go through a lengthy training program or wait for the passage of time until they have accrued sufficient field experience.

A solution to this problem is SeisTutor. The topic matter consists of the necessary seismic interpretation skills and the fundamentals of this sector. This formerly tacit knowledge has been made explicit and may now be taught to others. Each learner's competency level and learning style is unique, so SeisTutor was built to the first question, judge, and deliver instruction through an exclusive and tailored learning plan known as "tutoring strategy." The student's performance during tutoring sessions is also evaluated.

Even though SeisTutor does not promise comprehensive mastery of the subject matter, it is a modest attempt to make available the knowledge on this uncommon domain as a tutor-able and learner-centric form. SeisTutor (as per individual learner preferences).

Listed below are some of the book's notable contributions to the field of intelligent tutoring.

1.6.1 Development of Adaptive Knowledge Base

Through this book, we have constructed the adaptable domain model, which considers each learner's characteristics and performance.

Seismic data interpretation is covered in detail in the knowledge capsules provided by SeisTutor. The primary purpose is to collect tacit information from experts and turn it into an explicit form for dissemination. Techniques for tactic knowledge acquisition are already in existence. To take it a step further, this explicit information is organized into twelve different pedagogy styles, each with its own set of characteristics (a mix of the learner profile and each student's learning style). SeisTutor's adaptable knowledge base enables it to give material suited to each learner's specific needs, which is vital because each learner has a unique learning profile and style. Consequently, SeisTutor contains twelve knowledge capsules, allowing it to accommodate a diverse spectrum of pupils.

As an illustration, consider the following scenario in which pedagogical styles are utilized to offer customized learning materials.

Beginner 9, Intermediate 4, and Expert 7 are the learner's level scores, while Imagistic = 9, Acoustic = 3, Intuitive = 5, Active = 8. The learner's level scores are Beginner 9, Intermediate 4, and Expert 7. Using a weighted average of student performance on both tests, SeisTutor maintains a priority queue for instructional approaches evaluated in order of importance. These educational strategies are ranked in order of importance as follows: In this case, the (Imagistic-Beginner, 1), (Intuitive Beginner, 2), and (Acoustic Beginner, 4) are all appropriate. The next combinations of learning styles and learning levels scores are likewise listed and preserved after the highest scores of "Imagistic" and "Beginner," in the same way as the highest scores of "Imagistic" and "Beginner" are kept after the highest scores of "Beginner." Learning style preferences for the learner in this pretest example are "Beginner" and "Imagistic," respectively, and the pedagogy style "Beginner + Imagistic" is employed for this particular individual.

1.6.2 Learner-Centric Curriculum Recommendation

The purpose of this chapter's contribution is to demonstrate how to use the adaptive domain and pedagogy models to develop a learner-centric curriculum that considers each learner's unique characteristics and skills. ITS makes extensive use of the pedagogy model to construct a learning path that is adaptive and tailored to the needs of each individual. By combining the best features of both models, this ITS provides students with a wide range of alternatives for their educational travels. To provide customized education, it is necessary to assess the student's prior knowledge. The customized curriculum was developed due to the use of the bug model. By employing the bug model, it is possible to uncover bugs (misconceptions) during the pretest. When recommending post-test corrective steps, this model is effective in the research.

A few parallels can be drawn between this study and the bug behavior model. At the outset of a learning session, however, it is used to detect the learner's existing knowledge and errors and prescribe a specific curriculum plan for the remainder of the session.

1.6.3 Personalized Tutoring Strategy

The suggested ITS incorporates intelligence techniques, which will be discussed in further detail in the following sections, to provide a personalized learning environment for the student. As part of an exclusive "tutoring strategy," these intelligence techniques take into account a variety of learner characteristics (learner level and learning style) as well as educational parameters (previous learner knowledge, facial expressions during learning, learner performances) to provide a personalized learning path (learner-centric curriculum) to the learner as part of an exclusive "tutoring strategy."

1.6.4 Identification of Learner Understand-Ability

Learners are assessed on their ability to summarize and write about their understanding of a concept using a performance analyzer test explicitly developed. Based on this information, the rating for "Degree of Understanding" is computed. It is also possible to quantify a learner's progress on the route to lifelong learning by administering periodic performance evaluations. This is a non-psychological examination of the students present in the class.

1.6.5 Identification of Learner Emotional State

It is being developed as an Emotion Recognition Module (CNN-based Emotion Recognition Module) that uses machine learning. This module assists learners in identifying their feelings as they relate to their present course of the investigation. This is a psychological assessment of the student's mental well-being.

The literature review findings indicate that students' emotions can be used to evaluate their inner cognitive state and the performance of the tutoring system when assessing their internal cognitive state (learning material and tutoring mechanism).

1.7   Organization of Content

This book is organized according to the traditional component architecture to provide readers with a comprehensive overview of the ITS domain. The book is broken up into eight sections. The modeling of the domain and the pedagogy are the focus of the first three chapters, while the construction of the ITS is the subject of the following two chapters. ITS intelligence features are discussed in detail in each of these chapters. In the final three chapters, we'll look at how an ITS with built-in intelligence functions in practice.

There has been a lot of interest in acquiring and representing a domain knowledge model in the AI domains. A history of ITS domain acquisition concerns, going back to the sector's inception, is presented in Chap. 2. Various tactics and procedures are examined in this chapter. Domain knowledge engineering raises an important epistemological question addressed in the first section of the paper. Consideration is then given to various knowledge representation languages, ranging from expressiveness to inferential power to cognition plausibility to pedagogical emphasis. Toward the end of the chapter, connections are made to the following chapters in this book section.

Chapter 3 provides an overview of the pedagogical model research that has been done. This model is the brain of an ITS because it has the cognitive capacity that mirrors human intellect. Several studies on learner route sequencing are briefly discussed, considering their value, usability, and importance of recording psychological aspects during tutoring. There is a strong connection between this chapter and other sections inside the book.

A complete study of adaptive knowledge bases and hierarchical representations of knowledge is covered in Chap. 4, which explains the construction of the domain model. There is a strong connection between this chapter and other sections inside the book.

This chapter discusses the Pedagogy model's ITS and adaptability modules, the customized curriculum sequencing model, the tutoring strategy recommendation model, and the learner performance analysis module (psychological and non-psychological). There is a strong connection between this chapter and other sections inside the book.

ITS adaptive model implementation is discussed in Chap. 6. The performance of ITS is demonstrated with examples in this chapter. Toward the end of the chapter, connections are made to the following chapters in this book section.

To measure an ITS's effectiveness, performance measures are employed in Chap. 7. Evaluating an ITS's efficiency is a critical first step in determining not just the student's performance but also the platform's productivity.

In Chap. 8, the study's findings are based on performance metrics, and the conclusions are taken.

References

1. Nwana, H.S.: Intelligent tutoring systems: an overview. Artif. Intell. Rev. **4**, 251–277 (1990)
2. Wenger, E.: Artificial Intelligence and Tutoring Systems: Computational Approaches to the Communication of Knowledge. University of California, Irvine (1986)
3. Baffes, P., Mooney, R.: Refinement-Based student modeling and automated bug library construction. J. Artif. Intell. Educ. **7**(1), 75–116 (1996)
4. Chou, C.Y., Chan, T.W., Lin, C.J.: Redefining the learning companion: the past, present, and future of educational agents. Comput. Educ. **40**(3), 255–269 (2003)
5. Conati, C., Gertner, A.S., VanLehn, K., Druzdzel, M.J.: On-line student modeling for coached problem solving using Bayesian networks. In: User Modeling, pp. 231–242. Springer, Vienna (1997)
6. Zapata-Rivera, D., Greer, J.: Interacting with inspectable bayesian student models. Int. J. Artif. Intell. Educ. **14**, 127–168 (2004)
7. Kavcic, A., Pedraza-Jiménez, R., Molina-Bulla, H., Valverde-Albacete, F. J., Cid-Sueiro, J., Navia-Vazquez, A.: Student modeling based on fuzzy inference mechanisms. In: The IEEE Region 8 EUROCON 2003. Computer as a Tool, Vol. 2, pp. 379–383. IEEE (2003)
8. Katsionis, G., Virvou, M.: A cognitive theory for affective user modelling in a virtual reality educational game. In: Proceedings of 2004 IEEE International Conference on Systems, Man, and Cybernetics, pp. 1209–1213 (2004)
9. Evens, M.W., Brandle, S., Chang, R., Freedman, R., Glass, M., Lee, Y.H., Shim, L.S., Woo, C.W., Zhang, Y., Zhou, Y., Michael, J.A., Rovick, A.A.: CIRCSIM-Tutor: an intelligent tutoring system using natural language dialogue. In: Proceeding of Twelfth Midwest AI and Cognitive Science Conference, Oxford, pp. 16–23 (2001)
10. Vicari, R., Flores, C.D., Seixas, L., Gluz, J.C., Coelho, H.: AMPLIA: a probabilistic learning environment. Int. J. Artif. Intell. Educ. **18**(4), 347–373 (2008)
11. Mitrovic, A.: An intelligent SQL tutor on the web. Int. J. Artif. Intell. Educ. **13**(2–4), 173–197 (2003)
12. Khuwaja, A. Ramzan, M.W. Evens, J.A. Michael, A.A. Rovick, Architecture of CIRCSIM-Tutor (v.3): a smart cardiovascular physiology tutor. In: Proceedings of the 7th Annual IEEE Computer-Based Medical Systems Symposium, Winston-Salem, NC, pp. 158–163 (1994)
13. Naser, S.S.A.: JEE-Tutor: an intelligent tutoring system for java expressions evaluation. Inf. Technol. J. **7**(3), 528–532 (2008)
14. Wolcott, L.: The distance teacher as reflective practitioner. Educ. Technol. **1**, 39–43 (1995)
15. Chien, T.C., Yunus, M., Suraya, A., Ali, W.Z.W., Bakar, A.: The effect of an intelligent tutoring system (ITS) on student achievement in algebraic expression. Online Submission **1**(2), 25–38 (2008)
16. Gaudioso, E., Montero, M., Hernandez-Del-Olmo, F.: Supporting teachers in adaptive educational systems throughvpredictive models: a proof of concept. Expert Syst. Appl. **39**(1), 621–625 (2012)
17. Virvou, M., Kabassi, K.: F-smile: an intelligent multi-agent learning environment. In: Proceedings of 2002 IEEE International Conference on Advanced Learning Technologies-ICALT, pp. 144–149. Citeseer (2002)
18. Kearsley, G.: Online learning: personal reflections on the transformation of education. Educ. Technol. (2005)
19. Skinner, B.F.: Teaching machines. Science **128**
20. Uhr, L.: Teaching machine programs that generate problems as a function of interaction with students. In: Proceedings of the 1969 24th National Conference, pp. 125–134 (1969)
21. Sleeman, D., Brown, J.S.: Introduction: intelligent tutoring systems. In: Intelligent Tutoring Systems, pp. 1–11. Academic Press (1982)
22. Woods, P., Hartley, J.R.: Some learning models for arithmetic tasks and their use in computer based learning. Br. J. Educ. Psychol. **41**(1), 38–48 (1971)
23. Wang, T., Mitrovic, A.: Using neural networks to predict student's performance. In: Proceeding of International Conference on Computers in Education, pp. 969–973 (2002)

24. Suppes, P.: On using computers to individualize instruction. Comput. Am. Educ. 11–24 (1967)
25. Moritz, S., Blank, G.: Generating and evaluating object-oriented designs for instructors and novice students. In: Intelligent Tutoring Systems for Ill-Defined Domains: Assessment and Feedback in Ill-Defined Domains, p. 35 (2008)
26. Gertner, A.S., VanLehn, K.: Andes: a coached problem solving environment for physics. In: Proceeding of 5th International Conference, pp. 133–142. Intelligent Tutoring Systems, Berlin (2000)
27. Chakraborty, S., Roy, D., Basu, A.: Development of knowledge based intelligent tutoring system. Adv. Knowl. Based Syst. Model Appl. Res. **1**, 74–100 (2010)
28. Self, J.: Theoretical foundations for intelligent tutoring systems. J. Artif. Intell. Educ. **1**(4), 3–14 (1990)
29. Freedman, R., Ali, S.S., McRoy, S.: Links: what is an intelligent tutoring system? Intelligence **11**, 15–16 (2000)
30. Anania, J.: The effects of quality of instruction on the cognitive and affective learning of students. University of Chicago (1981)
31. The influence of instructional conditions on student learning and achievement. Eval. Educ. **7**(1), 1–92 (1983)
32. Bloom, B.S.: The 2 sigma problem: the search for methods of group instruction as effective as one-to-one tutoring. Educ. Res. **13**(6), 4–16 (1984)
33. Merrill, D.C., Reiser, B.J., Ranney, M., Trafton, J.G.: Effective tutoring techniques: a comparison of human tutors and intelligent tutoring systems. J. Learn. Sci. **2**(3), 277–305 (1992)
34. VanLehn, K.: The relative effectiveness of human tutoring, intelligent tutoring systems, and other tutoring systems. Educ. Psychol. **46**(4), 197–221 (2011)
35. Steenbergen-Hu, S., Cooper, H.: A meta-analysis of the effectiveness of intelligent tutoring systems on college students' academic learning. J. Educ. Psychol. **106**(2), 331
36. Ma, W., Adesope, O.O., Nesbit, J.C., Liu, Q.: Intelligent tutoring systems and learning outcomes: a meta-analysis. J. Educ. Psychol. **106**(4), 901 (2014)
37. Kulik, J.A., Fletcher, J.: Effectiveness of intelligent tutoring systems a meta-analytic review. Rev. Educ. Res. **106**, 0034654315581420 (2015)
38. Sierra, E., García-Martínez, R., Cataldi, Z., Britos, P., Hossian, A.: Towards a methodology for the design of intelligent tutoring systems. Res. Comput. Sci. J. **20**, 181–189 (2006)
39. Chrysafiadi, K., Virvou, M.: Student modeling approaches: a literature review for the last decade. Expert Syst. Appl. **40**(11), 4715–4729 (2013)
40. Aleven, V., Mclaren, B., Roll, I., Koedinger, K.: Toward meta-cognitive tutoring: a model of help seeking with a cognitive tutor. Int. J. Artif. Intell. Ed. **16**(2), 101–128 (2006)
41. Carmona, C., Conejo, R.: A learner model in a distributed environment. In: Adaptive Hypermedia and Adaptive Web-Based Systems, pp. 353–359. Springer (2004)
42. Trella, M., Conejo, R., Bueno, D., Guzmán, E.: An autonomous component architecture to develop WWW-ITS. In: Proceedings of the Workshops on Adaptive Systems for Web-Based Education, pp. 69–80. Malaga (2002)
43. Lu, C.-H., Wu, C.-W., Wu, S.-H., Chiou, G.-F., Hsu, W.-L.: Ontological support in modeling learners' problem solving process. J. Educ. Technol. Soc. **8**(4), 64–74 (2005)
44. Kumar, A.N.: Using enhanced concept map for student modeling in programming tutors. In: FLAIRS Conference, 2006, pp. 527–532.
45. Glushkova, T.: Adaptive model for user knowledge in the e-learning system. In: Proceedings of the 9th International Conference on Computer Systems and Technologies and Workshop for PhD Students in Computing, p. 78. ACM (2008)
46. Limongelli, C., Sciarrone, F., Temperini, M., Vaste, G.: Adaptive learning with the LS-plan system: a field evaluation. Learn. Technol. IEEE Trans. **2**(3), 203–215 (2009)
47. Rich, E.: Stereotypes and user modelling. In: Kobsa, A., Wahlster, W. (eds.) User Models in Dialog Systems. Symbolic Computation, pp. 35–51. Springer, Berlin Heidelberg (1989). https://doi.org/10.1007/978-3-642-83230-7_2

48. Virvou, M., Du Boulay, B.: Human plausible reasoning for intelligent help. User Model. User-Adap. Inter. **9**(4), 321–375 (1999)
49. Kay, J.: Stereotypes, student models and scrutability. In: Gauthier, G., Frasson, C., VanLehn, K. (eds.) Intelligent Tutoring Systems. Lecture Notes in Computer Science, vol. 1839, pp. 19–30. Springer, Berlin Heidelberg (2000). https://doi.org/10.1007/3-540-45108-0_5
50. Grigoriadou, M., Kornilakis, H., Papanikolaou, K.A., Magoulas, G.D.: Fuzzy inference for student diagnosis in adaptive educational hypermedia. In: Vlahavas, I.P., Spyropoulos, C.D. (eds.) Methods and Applications of Artificial Intelligence. Lecture Notes in Computer Science, vol. 2308, pp. 191–202. Springer, Berlin Heidelberg (2002). https://doi.org/10.1007/3-540-46014-4_18.
51. Tsiriga, V., Virvou, M.: Evaluation of an intelligent web-based language tutor. In: Palade, V., Howlett, R.J., Jain, L. (eds.) Knowledge-Based Intelligent Information and Engineering Systems. Lecture Notes in Computer Science, vol. 2774, pp. 275–281. Springer, Berlin Heidelberg (2003). https://doi.org/10.1007/978-3-540-45226-3_38
52. Martins, A.C., Faria, L., De Carvalho, C.V., Carrapatoso, E.: User modeling in adaptive hypermedia educational systems. J. Educ. Technol. Soc. **11**(1), 194–207 (2008)
53. Faraco, R.A., Rosatelli, M.C., Gauthier, F.A.: An approach of student modelling in a learning companion system. In: Advances in Artificial Intelligence–IBERAMIA 2004, pp. 891–900. Springer (2004)
54. Surjono, H.D., Maltby, J.R.: Adaptive educational hypermedia based on multiple student characteristics. In: Advances in Web-Based Learning-ICWL 2003, pp. 442–449. Springer (2003)
55. Mitrovic, A.: Fifteen years of constraint-based tutors: what we have achieved and where we are going. User Model. User-Adap. Inter. **22**(1–2), 39–72 (2012)
56. Mitrovic, A., Mayo, M., Suraweera, P., Martin, B.: Constraint-based tutors: a success story. In: Monostori, L., Váancza, J., Ali, M. (eds.) Engineering of Intelligent Systems. Lecture Notes in Computer Science, vol. 2070, pp. 931–940. Springer, Berlin Heidelberg (2001). https://doi.org/10.1007/3-540-45517-5_103
57. Mitrovic, A.: An intelligent SQL tutor on the web. Int. J. Artif. Intell. Educ. (IJAIED) **13**, 173–197 (2003)
58. Le, N.-T., Menzel, W.: Using Weighted constraints to diagnose errors in logic programming—the case of an ill-defined domain. Int. J. Artif. Intell. Ed. **19**(4), 381–400 (2009)
59. Collins, A., Burstein, M., Baker, M.: Human plausible reasoning. DTIC Document Tech. Rep. (1988)
60. Gwo-Hshiung, T.: Multiple attribute decision making: methods and applications. (2010)
61. Millán, E., Loboda, T., Pérez-de-la Cruz, J.L.: Bayesian networks for student model engineering. Comput. Educ. **55**(4), 1663–1683 (2010)
62. Jongsawat, N., Tungkasthan, A., Premchaiswadi, W.: Dynamic data feed to Bayesian network model and SMILE web application. In: Rabai, A. (ed.) Bayesian Network, pp. 155–166 (2010)
63. SMILE, M.J.: Structural modeling, inference, and learning engine and genie: a development environment for graphical decision-theoretic models. In: Proceedings of the Sixteenth National Conference on Artificial Intelligence (AAAI-99), pp. 18–22 (1999)
64. Gertner, A.S., Conati, C., VanLehn, K.: Procedural help in Andes: generating hints using a bayesian network student model. In: Proceedings of the Fifteenth National/Tenth Conference on Artificial Intelligence/Innovative Applications of Artificial Intelligence, pp. 106–111. AAAI '98/IAAI '98. American Association for Artificial Intelligence (1998)
65. Millán, E., Pérez-De-La-Cruz, J.L.: A Bayesian diagnostic algorithm for student modeling and its evaluation. User Model. User-Adap. Inter. **12**(2–3), 281–330 (2002)
66. Bunt, A., Conati, C.: Probabilistic student modelling to improve exploratory behaviour. User Model. User-Adap. Inter. **13**(3), 269–309 (2003)
67. Zapata-Rivera, J.-D.: Indirectly visible bayesian student models. In: BMA (2007)
68. Schiaffino, S., Garcia, P., Amandi, A.: Eteacher: providing personalized assistance to e-learning students. Comput. Educ. **51**(4), 1744–1754 (2008)

69. Conati, C., Maclaren, H.: Empirically building and evaluating a probabilistic model of user affect. User Model. User-Adap. Inter. **19**(3), 267–303 (2009)
70. MuÃśoz, K., Kevitt, P.M., Lunney, T., Noguez, J., Neri, L.: PlayPhysics: an emotional games learning environment for teaching physics. In: Bi, Y., Williams, M.-A. (eds.) Knowledge Science, Engineering and Management. Lecture Notes in Computer Science, vol. 6291, pp. 400–411. Springer, Berlin Heidelberg (2010). https://doi.org/10.1007/978-3-642-15280-1_37
71. Chieu, V.M., Luengo, V., Vadcard, L., Tonetti, J.: Student modeling in orthopedic surgery training: exploiting symbiosis between temporal Bayesian networks and fine-grained didactic analysis. Int. J. Artif. Intell. Educ. **20**(3), 269–301 (2010)
72. Sabourin, J., Mott, B., Lester, J.C.: Modeling learner affect with theoretically grounded dynamic bayesian networks. In: D'Mello, S., Graesser, A., Schuller, B., Martin, J.-C. (eds.) Affective Computing and Intelligent Interaction. Lecture Notes in Computer Science, vol. 6974, pp. 286–295. Springer, Berlin Heidelberg (2011). https://doi.org/10.1007/978-3-642-24600-5_32
73. Suraweera, P., Mitrovic, A.: KERMIT: a constraint-based tutor for database modelling. In: Intelligent Tutoring Systems. Lecture Notes in Computer Science, vol. 2363, pp. 377–387. Springer, Berlin Heidelberg (2002). https://doi.org/10.1007/3-540-47987-2_41
74. Bourdeau, J., Grandbastien, M.: Modeling tutoring knowledge. In: Nkambou, R., Bourdeau, J., Mizoguchi, R., (eds.) Advances in Intelligent Tutoring Systems. Studies in Computational Intelligence, vol. 308, pp. 123–143. Springer, Berlin Heidelberg (2010). https://doi.org/10.1007/978-3-642-14363-2_7
75. Riccucci, S.: Knowledge management in intelligent tutoring systems—AMS tesi di dottorato—AlmaDLuniversitÃă di bologna. (2008)
76. Bourdeau, J., Grandbastien, M.: Modeling tutoring knowledge. In: Nkambou, R., Bourdeau, J., Mizoguchi, R. (eds.) Advances in Intelligent Tutoring Systems. Studies in Computational Intelligence, vol. 308, pp. 123–143. Springer, Berlin Heidelberg (2010). https://doi.org/10.1007/978-3-642-14363-2_7
77. Jeremić, Z., Jovanović, J., Gašević, D.: Student modeling and assessment in intelligent tutoring of software patterns. Expert Syst. Appl. **39**(1), 210–222 (2012)
78. Lesta, L., Yacef, K.: An intelligent teaching assistant system for logic. In: Intelligent Tutoring Systems. Lecture Notes in Computer Science, vol. 2363, pp. 421–431. Springer, Berlin Heidelberg (2002). https://doi.org/10.1007/3-540-47987-2_45
79. Aleven, V.: Rule-based cognitive modeling for intelligent tutoring systems. In: Advances in Intelligent Tutoring Systems, pp. 33–62. Springer (2010)
80. Mitrovic, A., Martin, B: Evaluating the effectiveness of feedback in SQL-tutor. In: Proceedings International Workshop on Advanced Learning Technologies, IWALT, pp. 143–144 (2000)
81. Desmarais, M.C., Baker, R.S.: A review of recent advances in learner and skill modeling in intelligent learning environments. User Model. User-Adap. Inter. **22**(1–2), 9–38 (2012)
82. Butcher, K.R., Aleven, V.: Using student interactions to foster rule–diagram mapping during problem solving in an intelligent tutoring system. J. Educ. Psychol. **105**(4), 988 (2013)
83. Kodaganallur, V., Weitz, R.R., Rosenthal, D.: A comparison of model-tracing and constraint-based intelligent tutoring paradigms. Int. J. Artif. Intell. Educ. (IJAIED) **15**, 117–144 (2005)
84. Mitrovic, A., Koedinger, K.R., Martin, B.: A comparative analysis of cognitive tutoring and constraint-based modelling. In: Brusilovsky, P., Corbett, A., Rosis, F.D. (eds.) User Modeling. Lecture Notes in Computer Science, vol. 2702, pp. 313–322. Springer, Berlin Heidelberg (2003). https://doi.org/10.1007/3-540-44963-9_42
85. Nkambou, R.: Modeling the domain: an introduction to the expert module. In: Nkambou, R., Bourdeau, J., Mizoguchi, R. (eds.) Advances in Intelligent Tutoring Systems, vol. 308, pp. 15–32. Studies in Computational Intelligence. Springer, Berlin Heidelberg (2010). https://doi.org/10.1007/978-3-642-14363-2_2
86. Prabhu, M.G.H.K.L.V. (ed.): Handbook of Human-Computer Interaction, 2nd edn, pp. 1551–1582. North-Holland (1997)
87. Baschera, G.-M., Gross, M.: Poisson-based inference for perturbation models in adaptive spelling training. Int. J. Artif. Intell. Educ. **20**(4), 333–360 (2010)

88. Jaques, P.A., Seffrin, H., Rubi, G., de Morais, F., Ghilardi, C., Bittencourt, I.I., Isotani, S.: Rule-based expert systems to support step-by-step guidance in algebraic problem solving: the case of the tutor Pat2Math. Expert Syst. Appl. **40**(14), 5456–5465 (2013)
89. Corbett, A.T., Anderson, J.R.: Knowledge tracing: modeling the acquisition of procedural knowledge. User Model. User-Adap. Inter. **4**(4), 253–278 (1994)
90. Cen, H., Koedinger, K., Junker, B.: Learning factors analysis—a general method for cognitive model evaluation and improvement. In: International Conference on Intelligent Tutoring Systems, pp. 164–175. Springer (2006)
91. Pavlik Jr., P.I., Cen, H., Koedinger, K.R.: Performance factors analysis—a new alternative to knowledge tracing. Online Submission (2009)
92. Blessing, S.B., Gilbert, S.B., Ourada, S., Ritter, S.: Authoring model-tracing cognitive tutors. Int. J. Artif. Intell. Educ. **19**(2), 189 (2009)
93. Nkambou, R., Bourdeau, J., Psyché, V.: Building intelligent tutoring systems: an overview. In: Advances in Intelligent Tutoring Systems, pp. 361–375. Springer (2010)
94. Paviotti, G., Rossi, P.G., Zarka, D.: Intelligent tutoring systems: an overview. Pensa Multimedia (2012)
95. P. Fournier-Viger, R. Nkambou, and E. M. Nguifo, "Building intelligent tutoring systems for ill-defined domains," in Advances in intelligent tutoring systems. Springer, 2010, pp. 81–101.
96. Clancey, W.J.: Use of Mycin's rules for tutoring. Rule-Based Expert Syst. Addison-Wesley Reading 19–2 (1984)
97. Nye, B.D., Graesser, A.C., Hu, X.: AutoTutor and family: a review of 17 years of natural language tutoring. Int. J. Artif. Intell. Educ. **24**(4), 427–469 (2014)

Chapter 2
Domain Modeling

2.1 Introduction

Students can benefit from intelligent tutoring systems (ITSs), which are computer programs meant to aid them in learning and enhance their abilities in a given subject area. These tutoring systems need to have an explicit representation of domain knowledge to provide successful services. For the system to use the representation to solve issues in the domain successfully, the system must be provided with mechanisms that allow the system to do so efficiently and effectively. Many Artificial Intelligence research programs have been launched since the field's inception, all of which have sought to address the tough task known as acquiring and modeling the domain-related knowledge and model. You will need to be creative [1–3].

To be effective, systems based on knowledge, and in particular systems based on expert knowledge, must explicitly reflect this domain's competency to be effective. Similarly, to facilitate the dissemination [4, 5] and the acquisition of the related knowledge, an expert module that can generate and answer domain problems and provide learners with access to such information is required. Academics in the field of information technology should focus their efforts on building the explicit model with respect to related knowledge that incorporates robust perceptive procedures as a result of this finding. It is recommended that the expert module be used in an instructional technology system to interpret the behavior of pupils [6]. The formalisms used to define and apply domain knowledge must be taken into consideration not only in terms of their nature and utility, but they must also be taken into consideration in terms of how they are utilized.

Various methods have been created to communicate subject expertise clearly and succinctly. Solution finding has been accomplished through the application of a wide range of disciplines to the problems that have been identified. In contrast to the epistemological approach taken by philosophers, psychologists, and education scientists such as Piaget et al. [7], Gagné et al. [8], Flores and Winograd [9], Twitchell and Merrill [10], and others, the fields of artificial intelligence and cognitive science provide solutions that make it easier to express and implement knowledge computationally. Anderson [11], Collins and Quillian [12], Minsky [13], Sowa [14] are examples of authors who have written about this topic. Many approaches are available to meet the AIED's needs for a sound knowledge of the representation of models and inferential procedures linked to information technology systems. These approaches should be implemented into the system.

2.2 An Epistemological Outlook Related to Domain Knowledge

The study of epistemology is concerned with how we come to know what we do. According to Piaget's idea of epistemology, "epistemology" is used to describe the study of knowledge (1997). The epistemic inquiry includes questions about the nature of knowledge (gnoseological issues), the methods employed to produce it, and questions about the information's value and validity. These considerations, in our view, are critical to the formalization of knowledge in a particular area. While discussing the personal knowledge, behavior, and inference tools available within a specific field, epistemological issues are essential. With this approach, we can call into demand numerous parts of knowledge, including the creation modes, underlying basics, and production dynamics. As evidenced by the research of [15] while developing knowledge-based systems, this epistemological perspective is taken into consideration (1992).

In the conventional approaches of knowledge engineering that are still in use today, the epistemological and computational levels are kept separate from one another. As a starting point, it is critical to analyze the constraints imposed by domain knowledge's conceptual structure and the most prevalent approaches to concluding patterns in domain knowledge. The next stage is to implement the procedures and formalizations that will allow these constituents to be formally observed in the future. Ontology and inference model on concept-based system necessity be specified on an epistemological level before they can be implemented. For example, ontology is a conceptual model related to the field of knowledge which emphasizes the existence of knowledge based on the nature of the knowledge domain itself. Inference modeling can be used to solve or complete a job by managing ontology, and the inference model demonstrates nature along the inference structure needed to do so.

In developing an ITS, it is critical to consider the type of information that will be taught or learned. The definition of the behavior of knowledge engaged with the acquiring procedure is critical in epistemology since it is a component of the learning process. Declarative knowledge and procedural knowledge are the two most fundamental types of domain knowledge described by the most generally used classification system in the fields of artificial intelligence and psychology, respectively. Gagne [8] and Bloom [16] were the first to categorize knowledge and abilities exactly in educational psychology. These contend that various classifications of knowledge need diverse techniques of educating and learning. Modern cognitive theory enabled the development of new knowledge-typing systems that were better suited to computer representations in the future. Merrill [17] provides an example of this by arranging information in a matrix by placing content types (such as facts, concepts, and processes) on one axis and performance levels (such as recall, apply, and create) on the other. Merrill's Component Display Theory is an example of this. Compared to hierarchical representations, such as those provided by Gagné, this matrix technique is more expressive and straightforward. While epistemology is concerned with identifying distinct types of information, it is also concerned with the qualities and functions of these various types of knowledge in the solution of issues.

A more difficult multidimensional model, proposed by Kyllonen et al. [18], makes distinctions between different types of knowledge by employing a hierarchy based on cognitive complexity to discern between them. The value of autonomy and the processing speed are required to execute the work; these types of knowledge are organized in a certain way. The authors of [19] argue that, while exact classification stays vital, the extra pragmatic typology field's knowledge would be developed that considers the environment in which the information is employed. An epistemological study of the knowledge that generates required ontological categories and the excellence of knowledge needed was offered, and it was advocated that this study be done from the position of "knowledge-in-use." This perspective on domain knowledge has received very little attention in the educational literature because it is only infrequently explored in relation to ITSs (Information Technology Systems) (see the preceding paragraphs). Most information technology systems are focused on procedural domains with limited scope, which are focused on the concepts and tasks required by those domains, which could be a contributing factor.

Nonetheless, the rising focus has remained on the crucial role of epistemology in the learning sciences, which is a positive development. It has been demonstrated that instructors' specific notions about the nature of information and learning impact the curriculum, pedagogy, and assessment decisions that are made in their courses [20, 21]. An epistemological perception of field knowledge has taken a component of recognized methodology associated with knowledge-dependent structures for quite some time now. When it comes to knowledge-based systems, one of their most prevalent aspects is the epistemological investigation of the system's dual most essential parts (knowledge around the field and information toward inference methods required to answer the problem) [15]. It delivers the background of knowledge-structured primitives that characterize many kinds of perceptions, information generators, and organizational relationships like inheritance and various categories related to problem-solving methods, such as hypothesis testing, among other things [22, 23]. In the same manner that a knowledge-based system is equivalent to another knowledge-based approach, the skilled component of the ITSs stands analogous to another expert module. In the following section, we will go through several computational ways of demonstrating and reasoning about knowledge. We will draw a close to our investigation of epistemological considerations at this point because the essential objective is not the proposal of new practices in the domain of information technology systems.

2.3   Preliminary Research on Domain Model in IT

The ITS/E-learning systems' adaptability is built on the domain model. Seismic data interpretation is regarded as a subject domain for the time being. In the absence of standards for seismic interpretation, the subject of "seismic data interpretation" (SDI) is found to have a high degree of tacitness. For this reason, knowledge received by first-hand experience is referred to as "experiential knowledge." To build an adaptable domain model, it was necessary to obtain tacit knowledge and then realign that information per learner profiles and learning styles.

2.3.1   The Black Box Models

This section takes an entirely different approach to describe problem states than the technique taken by the learner in the previous section. In this context, the SOPHIE I [24] black-box model system is utilized as a classic example of electrical troubleshooting to demonstrate its application. Because it is the black-box model, it stands impossible to discern what is going on within [25]. When students are working through the many obstacles that can emerge with a circuit, they can use an expert system to analyze their measurements. Using a set of equations as a foundation, the expert system mimics a troubleshooting model to assist problem-solving. The instructor will use these equations to arrive at conclusions about what to do next and how to proceed with the assignment. Although students are expected to produce an explanation of the problem-solving framework as well as a rationale for selecting the appropriate technique, they ultimately should do so.

2.3.2   The Glass Box Models

The said glass box model exists as a transitional model which uses the same field constructs as the human professional. The model uses a different set of facts to support its claims instead of the model. The GUIDON [26], a medical diagnosis instructing system, is an outstanding example of a system designed in the manner of the glass box approach. The MYCIN platform, an expert bacterial infection treatment platform, served as the foundation for this system. A collection of hundreds of "if–then" procedures that link illness positions to identify in a probabilistic manner is contained within MYCIN. Their reasoning is based on the same symptoms. It represents the mind as doctors' reasoning, but they utilize a fundamentally different control structure to reach their conclusions, such as an exhaustive backward search for all probable causes. Those from the student are matched to questions that MYCIN would have asked, and GUIDON provides feedback based on the comparison.

2.3.3 The Cognitive Models

The cognitive models are intended to correspond to the human mind in representational formats and inference processes. Detecting cognitive faithfulness in the context of a domain knowledge module was regarded as some of the most important early achievements in intelligent instructing at the time of its discovery. When presented with problem-solving situations, ITSs are expected to behave the same manner as a student would. Computational methods are intended to construct cognition models of domain knowledge that correspond to how knowledge is encoded in a person's mental model [26]. Unlike the other approaches geared toward logical reasoning, this one is geared toward promoting cognitively reasonable thought. Briefly stated, the idea is to encode knowledge using the same representation type as the learner. Developing a system built on cognitive architecture, such as the ACT-R, is one method to take things ahead [27]. The ACT-R model of human cognition can help us better understand and imitate human understanding. The ACT-R architecture is an excellent choice for studying how humans see, think about, and interact with their surroundings. Several ITSs, including Algebra Tutor, Geometry Tutor, and LISP Tutor, were built in the early days of ACT-R (or ACT*, in its early form) production rules and the early days of ACT* production rules [27]. An analysis of human skill development revealed that instructors who reflect procedural domain knowledge are also cognitively oriented, as demonstrated by a study [28]. A similar tutoring method may be observed in the Sherlock Holmes series [29, 30].

2.4 Experiential (Tacit Domain Knowledge)

This section covers the knowledge and its forms in making it available for tutoring through the tutoring system.

Knowledge is seen as a critical asset. Knowledge management creates and disseminates knowledge and information and enables the effective use of knowledge to benefit the organization's strategic objectives [31]. According to Liebowitz and Beckman, knowledge management intrinsically maximizes an organization's knowledge-related effectiveness (by encouraging innovation, higher performance, and the development of new capabilities) [32]. According to [33], knowledge management is viewed as an organization's operational strategy since it enables the development of new innovative business processes that improve performance.

There are two types of knowledge: tacit and explicit knowledge. When explicit and tacit knowledge collaborate, a new original idea is born. Existing literature divides knowledge into theoretical and practical components, internal and external components, and foreground and background components. The most frequently seen element of knowledge is the categorization of explicit and tacit information [34, 35]. Tacit knowledge is experiencing knowledge that is distinct from explicit knowledge. Experiential knowledge is factual information received from personal experience with conditions in comparison with rational knowledge.

Tactic knowledge is critical for organizational growth since it consists of experiences, movement skills, implicit thumb rules, and intuition. There is a significant distinction between tacit and explicit knowledge. Explicit information is well-documented knowledge [36], whereas tacit knowledge lives in the minds of individuals as perspectives, unique ideas, and values [37, 38].

According to some experts, knowledge is neither entirely explicit nor entirely tacit. Knowledge, on the other hand, is somewhere in the middle. Nonaka [39] likewise emphasizes the relevance of links between explicit and tacit knowledge, stating that the two are not distinct entities but complement one another.

According to Polanyi, tacit knowledge is the knowledge that cannot be expressed verbally; it exists in the human brain. Polanyi [40] said succinctly, "we used to express less than we knew." Tacit knowledge is fundamentally distinct from explicit information in that it is more challenging to communicate and code than explicit knowledge. In the referenced article [41], author describe tacit knolwedge as "knowing how" and "knowing what", whereas "Knowing how" denotes something accessible by action, but "Knowing what" indicates something lucid. Tacit knowledge is easily comprehended using the proximal and distal concepts. The distal focuses on the move, whereas the proximal focuses on the specific activity, for instance, when preparing food. Proximal knowledge is about preparing food (focusing on the ingredients), whereas distal knowledge is about cooking food in general [42]. Tacit knowledge comprised a variety of sensory and mental data. In a way, an image is generated to make meaning [43]. Numerous definitions exist for tacit knowledge. This term, however, is utilized to distinguish tacit knowledge from explicit knowledge [44]. The management literature readily acknowledges that tacit knowledge is critical, difficult to replicate, distinctive, and acquired via experiences [45, 46].

2.5 Experiential Knowledge Acquisition Approaches

In this section, experiential knowledge is explained and made available for application is covered here. The advantages and disadvantages of the product are examined.

Organizational well-being relies heavily on employees gaining first-hand knowledge through work-related experiences. Acquiring tacit knowledge can be done in the following ways:

2.5.1 Cognitive Map

The subjects' beliefs are depicted using a cognitive map [47]. While researching the participants, Eden sought to understand and derive their perspectives and learn how they interpret and depict their immediate surroundings [48]. A map shows the subject's innate understanding of the world. The subject's thoughts on a specific issue are depicted in this map. Various types of cognitive maps are available [49]. This map style is known as a "cause and effect" map. Using graphical notation, the causal map can also be used to gather and describe tacit or experiential information (nodes indicate the opinion while edges indicate the linkage between the concepts).

2.5.2 Causal Map

Experiential knowledge can be explained using a causal map, which emphasizes the importance of a person's actions and provides supporting evidence for their beliefs. It is a graphical representation of experiential knowledge, where nodes represent the activity and edges reflect the relation or working processes (causalities). Some rules for beliefs that are constant for a particular task can be deduced by experiencing and mimicking specific experiential activities throughout time.

The subject's experiences are compiled in a causal map to gather experiential information. During map building, subjects are asked to depict the situation, limits, barriers, actions, behavior, and consequences. This kind of probing reveals previously unseen abilities. As a result, this shows how knowledge moves from being implicit to being explicit.

2.5.3 Self-Q

It is possible to learn a lot about someone by having them do a self-interview, or "Self-Q," on themselves. Subjects question themselves in this method. This is a promising strategy since the practitioner has a deep understanding of themselves that no one else has. Based on their knowledge, this tacit knowledge can only be expressed through a specific set of questions that the issues themselves devise [50].

2.5.4 Semi-Structured

The goal is predetermined in this knowledge extraction technique. The interviewer's job as a researcher is to get to know a subject's experience from beginning to end. Narrating one's own life experiences is deemed sufficient by Ambrosini and Bowman

[51] to make one's implicit knowledge explicit; hence, it is necessary to urge participants to do so. Narrating stories has revealed more information than simply providing facts and figures [52].

2.6 Ontology Engineering

Defining the term "ontology" is necessary before getting into the specifics. An ontology is a systematic explanation of a method of thinking (in this case, a domain) that comprises class, object, attribute, connection, and axiom definitions. In other words, an ontology is purely descriptive of a field. Ontologies are structured to allow automatic inference and are written in formal languages such as RDF or OWL. Ontologies are often constructed based on the beliefs held by the vast majority of members of a community. Ontologies specific to domains are a topic that particularly piques our interest here. As was said before, the idea of a domain ontology, as the eLearning community understands it, is a relatively recent development in the subject of ITS. However, domain ontology engineering is a rapidly developing field that is drawing a lot of attention from researchers in various disciplines. In addition to that, it is the basis upon which the Semantic Web is built. The study of strategies and applications that can effectively manage an ontology over its lifetime is known as ontology engineering. It needs a general, domain-independent method for building, refining, and evaluating ontologies [53].

The main parts of the ontology life cycle are the specification stage, the formalization stage, the maintenance stage, and the evaluation stage.

The specification stage defines the ontology's goal and scope. For the bulk of the time, domain experts are required, and the competence questions that the ontology should answer must always be defined. It is also influenced by how the ontology will be used. The formalization step creates a conceptual and formal model that fits the requirements of the specification stage. The ontology's upgrades and changes are monitored and validated during the maintenance stage. Finally, the evaluation stage looks at the completed ontology to see whether it meets the basic requirements and has the necessary features.

2.7 Building Domain Ontologies from Texts

This section is about understanding ontology using texts. The expert system should build a corpus of information about the topic of interest throughout the specification stage [54]. Of course, these corpora should be carefully selected and accurately represent the subject. Before one can master a domain ontology, the first task is to comprehend concepts, taxonomy, conceptual links, characteristics, instances, and axioms. For example, Text-2-Onto [55] TEXCOMON [56, 57] OntoLearn [58] as well as OntoGen are examples of systems that can learn ontologies [58]. The following sections discuss the most current strategies for extracting knowledge employed in each ontology learning sub-tasks. These sub-tasks have a natural language processing (NLP) technique and a statistical and machine learning technique.

2.7.1 Concept Extraction

The first step in ontology engineering is determining what concepts are. Concepts can be thought of as complex mental things with several pieces. Determining which domain classes are the most significant is known as concept extraction. Concepts are extremely relevant terms in the subject when applying terminological methodologies. As [59] describes, these phrases are typically drawn from the corpus. They believe that a concept should be able to be expressed in words. It is not easy to distinguish between domain terms and other words in this scenario, primarily when statistical filtering is utilized. The identified terms, which might consist of one or more words, can then be thought of as distinct concepts or classes, or they can be placed in broad categories already in the thesauri and vocabularies. In this scenario, an idea may not be associated with a word in the corpus. Clustering and machine learning are two different approaches to learning semantic classifications.

2.7.1.1 NLP-Based Techniques

Words are considered potential concepts by NLP-based concept-learning algorithms. These processes depend on language expertise and use parsers and taggers to discern the grammatical roles of phrases or to discover language patterns. Surface analysis is performed by running a part-of-speech tagger across the corpora and searching for manually specified patterns [60, 61]. Deep analysis is performed by applying an NLP parser [56, 62].

2.7.1.2 Statistical and Machine Learning Techniques

NLP-based techniques are frequently used in combination with statistical filtering. In statistical procedures, all critical things in a field are considered feasible concepts, and the weight of a phrase must be defined quantitatively. The well-known TF*IDF [63] and the C-value/NC-value are two of these quantitative indicators [64]. Depending on the application, the metrics used may differ. Clustering methods based on [65] distributional hypothesis can also be used to build semantic classes [66, 67]. In this sense, an idea is a group of related words. According to Harris' theory, words used in analogous settings typically have similar meanings on which phrase space models are based [68]. Collocations [69], co-occurrences [70], and latent semantic analysis can be used to identify the similarity between two terms [71], for example, assign a feature vector to each word based on the context in which it appears. The features are particular dependencies and how frequently they exist in the corpus. The vectors are then used to group similar terms and identify how related separate terms can use measurements such as similarity measures [72, 73]. The algorithms of Formal Concept Analysis (as explained by [74]) and Latent Semantic Indexing are comparable [75]. These methods return attribute and value pairs that relate to concepts. In addition to NLP-based methods, statistical approaches can identify just relevant domain keywords by analyzing how terms are utilized in distinct corpora [76]. Velardi et al. [58] applies an additional method for linguistically analyzing WordNet glosses (textual descriptions) to extract meaningful information about a specific idea and improve its features. This analysis can assist in locating synonyms and similar words, as well as in defining an idea. Understanding concepts demands the recognition of abstract classes and the characterization of concepts by specifying their traits, sub-classes, and relationships.

2.7.2   Attribute Extraction

Since concepts possess distinct features, it is essential to determine what they are. In his 1992 ontology, Guarino distinguishes between characteristics that are connected and those that are unrelated. Relationship-related characteristics include qualities and functions. In contrast, non-relational qualities are things like components. Almuhareb and Poesio [77] developed a novel technique for defining characteristics as qualities, sections, related objects, actions, and related agents in the wake of [78] as well as [79]. In this chapter, "attributes" relate to data type features such as "id," "name," and so on, whereas "object properties" are regarded as "conceptual relationships" and are explored in "Extracting Conceptual Relationships."

2.7.2.1   NLP-Based Techniques

According to [80], the best technique to determine what qualities mean is to examine Wood's linguistic interpretation [81]: If it is possible to state that Y is an A of X, then Y is a value of the X attribute A. (or the A of X). If there is no Y, then A cannot be an attribute. So that this language interpretation makes sense, linguistic patterns for recognizing qualities are also given.

2.7.2.2   Statistical and Machine Learning Techniques

As stated previously, statistical filtering and machine learning are frequently employed with natural language processing approaches. Poesio and Almuhareb [80] proposed a supervised classifier for learning attributes using morphological data, an attribute model, a question model, and an attributive-usage model. These models distinguish between multiple kinds of quality by using a particular categorization system as the criterion. Web-based concept descriptions are utilized in the research that [82] has conducted. Ravi and Pasca came up with the idea of a poorly supervised classifier that can acquire knowledge of attributes and values using only a limited number of examples [83].

2.7.3 Taxonomy Extraction

One of the essential facets of knowledge engineering is organizing information into taxonomies that illustrate how various classes can be generalized or specialized. These connections make it possible for ideas to build atop one another and give computers the ability to make choices [84].

2.7.3.1 NLP-Based Techniques

Hearst argued that lexico-syntactic patterns are the most typical approach to identifying taxonomic relationships [85]. Pattern-based approaches search the text for various lexico-syntactic patterns that indicate a taxonomic relationship. Cimiano and Volker [86] Patterns are typically stated as sequences, but they are also expressed as dependencies [56, 67]. Because a domain corpus is small and hierarchical patterns are unusual in domain-specific corpora, numerous systems look for taxonomical linkages in dedicated resources like WordNet [87] or on the web. Snow [87], Cimiano and Staab [88], Maedche and Staab [89], and Etzioni et al. [90] argue that using dependency pathways and wordnet for training a classifier to discover hyponym (is-a) correlations from a text can lessen the difficulties of manually finding patterns. Analyzing the architecture of multiword words is another way of uncovering taxonomy relationships through language (noun phrases). For instance, there is a relationship between a phrase and the same phrase with an adjective added (e.g., an intelligent man is-a man). This method is used by many people [56, 58, 91].

2.7.3.2 Statistical and Machine Learning Techniques

The statistical and machine learning techniques for learning about taxonomies are based on Harris' distributional hypothesis, as are the techniques for learning about concepts. Taxonomies are retrieved from text using agglomerative methods, and cluster hierarchies are constructed. Maedche et al. [92] distinguishes between two techniques of hierarchical clustering: the bottom-up methodology, which starts with specific elements and clusters those which are most similar, as well as the top-down approach, which groups all objects. This method has been used in several papers, including [93–95] (1998). Cimiano and Staab [88] define t as a sub-class of t2 if it appears in all syntactic contexts shared by t2. The grammatical contexts are being used as extracted features, and they are compared using a measure of similarity. For example, [88] used a directed Jaccard coefficient to show that the relationship between the two terms is a (t, t2) by dividing the total number of common characteristics by the number of attributes in term t. Cimiano and Staab [88] also recommends that individuals use a variety of evidence sources and approaches to learning about hierarchical relationships. Similarly, [96] advocates incorporating new concepts into an existing taxonomy using unsupervised algorithms incorporating statistical and syntactic data.

2.7.4 Conceptual Relationship Extraction

Conceptual relations encompass all types of non-taxonomic interactions between concepts. General conceptual relationships comprise any marked connection between a source idea (the relation's domain) and a target concept (the range of the relation). Specific conceptual links include synonymy, membership, possession, attributes, and causation. The following sections will discuss the many approaches to discussing specific and broad relationships.

2.7.4.1 NLP-Based Techniques

Conceptual relationship extraction also is known as template filling, function labeling, occurrence extraction, and frame filling in the area of information extraction. It uses lexico-semantic lexicons like FrameNet [97] and VerbNet [98] to discover relationships and assign roles (such as Agent and Theme) to the relationship's arguments. ASIUM [95] is a frame-based method that employs cue words to help individuals understand the relationships between concepts. Pustejovsky's identification of Qualia structures [79, 99] demonstrated that these qualia structures could assist in identifying specific associations. They developed a variety of linguistic patterns that illustrate the various roles specified by Pustejovsky. There has been a great deal of research on leveraging language patterns to identify ontological linkages in text. Conceptual relationship extraction also is known as template filling, function labeling, occurrence extraction, and frame filling in the area of information extraction. It uses lexico-semantic lexicons like FrameNet [97] and VerbNet [98] to discover relationships and assign roles (such as Agent and Theme) to the relationship's arguments.

2.7.4.2 Statistical and Machine Learning Techniques

The majority of relation extraction effort is a combination of statistical analysis and various language analysis levels. For instance, [57] employs typed dependencies to learn about relationships and statistical data, which are then used to determine whether the relationships should be included in the ontology or not. The 2003 work of [100], which uses inductive logic programming, and the 2004 work of Yamada and Baldwin, which employs both lexico-syntactic patterns and a maximum entropy model classifier, are more examples of how machine learning can be used to learn about qualia structures. Cimiano and Wenderoth [101] developed an algorithm for creating a collection of hints for each qualia role. It requires downloading the snippets of the first ten Google results that match the supplied clues, tagging the downloaded snippets for part of speech, matching regular expressions that describe the desired qualia role, and then assigning a weight to each of the retrieved qualia elements. Association rule learning is an intriguing method for discovering unlabeled associations. This technique creates rules based on the frequency with which pieces of the corpus appear together. The Text-to-Onto system has begun using this technique [89]. However, these linkages will subsequently require manual labeling, which is not always an easy task for the ontology engineer.

2.7.5 Instance Extraction

OP, or Ontology Population (OP), is a task for classifying data in which examples of concepts described in an ontology are sought. Data mining commonly uses Named Entity Recognition (PER), which recognizes specific entities such as people, places, and organizations by their names. For instance, extraction, WEBKB [102], and Know-It-All are two systems [90]. NLP-based techniques can be used to populate an ontology in various ways. In the taxonomy extraction part, a pattern-based technique like the one outlined leverages Hearst patterns [57, 85, 90, 103] or the structure of words [58]. Finding "is-a" associations is the goal of these methods. Other methods of learning a language, such as establishing or memorizing rules, are based on this approach. It has been suggested, for example, that the definition of acquisition rules is activated when defined language tags exist, as in the work of Amardeilh et al. It is easier to find instances of these items because the ontology has been tagged with these tags [66].

2.7.5.1 Statistical and Machine Learning Techniques

To populate an ontology, one can use either supervised or unsupervised method [104, 105]. The vector-feature similarity between each idea c and a term was employed by Cimiano and Volker [86] to place them in a category t. This was a procedure that had only been loosely supervised. Cimiano and Volker found that syntactic aspects (such as word windows and dependencies) were more effective in their experiments. By comparing the feature vectors of an instance to the feature vectors of the idea, their algorithm created a concept, for instance. Tanev and Magnini employed dependency parse trees to extract syntactic information in their 2006 paper [105]. In order to learn the algorithm, all they needed was a list of words for each category they were interested in analyzing. As a result, supervised methods for populating ontologies are more accurate. The problem is that they have to put together a training set manually, which is not scalable [105]. Examples of the supervised technique are found in the work of Fleischman (2001) as well as [106]. To categorize Named Entities, they used a machine learning system. An additional training dataset is used by web->KB [102] in order to locate named entities. Hypertext regions have been marked up to show occurrences of classes and relationships, resulting in this data. The system determines how to classify any web page or chain of links based on the ontology and the training data.

2.7.6 Axioms Extraction

There are axioms in ontologies, declarations of necessary and sufficient criteria used to limit the information and draw conclusions about additional information [107]. When learning ontology, extracting axioms is one of the more complex tasks. Only a few systems have attempted to solve the challenge of extracting axioms. KIF HASTI is a system that converts natural language phrases with explicit axioms into the logically formatted axioms of KIF [108]. This is a new effort to put natural language phrases (definitions) into description logic axioms, LExO2 [109].

2.7.6.1 NLP-Based Techniques

Syntactic transformation of natural language definitions into description logic axioms is required for axiom extraction methods that use natural language [109]. This is based on the assumption that definitions exist. Additionally, Volker and colleagues (2008) focus on learning a disjointedness by using a lexico-syntactic pattern that is used to locate enumerations. These people believe that when a list of items is broken down into discrete categories, it indicates that they are distinct. Defining equivalent classes is made more accessible by a pattern proposed by Zouaq and Nkambou [57]. Two words that have an appositive association have the same meaning and are therefore similar in this pattern. Intriguing is the method of [67]. In order to discover close relationships, it makes use of pathways in dependency trees. For example, "X solves Y" or "Y is solved by X" can be an inverse property for these connections.

2.7.6.2 Statistical and Machine Learning Techniques

To our knowledge, relatively few approaches systematically use machine learning to teach axioms. Machine learning classification was used by Volker et al. [109] to determine whether any two classes did not fit together. Disjointedness can be learned using lexical and logical features that are automatically extracted. An ontology's structure, associated textual resources, and other data kinds are examined in this process. In the end, the features are utilized to create a model that can classify everything at once.

2.8 Summary

This chapter explored the work in the field of Intelligent Tutoring System (ITS)/E-learning System/Hypermedia System for the domain model. The domain model, a crucial component of the ITS, has been explored in detail. This chapter emphasized the experiential knowledge domain and its acquisition and explication techniques. Toward the concluding part of this chapter, ontological engineering in the text through machine learning and NLP is discussed.

References

1. Brachman, R.J., Levesque, H.J.: Knowledge Representation and Reasoning. Morgan Kaufmann, San Francisco (2004). ISBN 978-1-55860-932-7
2. Nawaz, M.S., Hassan, M, Shaukat, S.: Impact of knowledge management practices on firm performance: testing the mediation role of innovation in the manufacturing sector of Pakistan. Pak. J. Commer. Soc. Sci. 8(1), 99–111. Clancey, B.: Acquiring, representing and evaluating a competence model of diagnostioc strategy. Technical Report: CS-TR-85–1067, Stanford University (1985)
3. Russell, S.J., Norvig, P.: Artificial Intelligence: A Modern Approach, 3rd edn. Prentice Hall, Upper Saddle River (2009)
4. Woolf, B.: Building Intelligent Interactive Tutors: Student-centered Strategies for Revolutionizing E-learning. Morgan Kaufmann, San Francisco (2008)
5. Wenger, E.: Artificial Intelligence and Tutoring Systems: Computational and Cognitive Approaches to the Communication of Knowledge. Morgan Kaufmann Publishers Inc., Los Altos (1987)
6. Corbett, A.T., Koedinger, K.R., Anderson, J.R.: Intelligent tutoring systems. In: Helander, T.K., Landauer, P. (eds.) Handbook of Human-Computer Interaction. Elsevier Science, Amsterdam (1997)
7. Piaget, J.: The Principles of Genetic Epistemology. Routledge, London (1997)
8. Gagne, R.M.: The Conditions of Learning, 4th edn. Holt Rinehart and Winston, New York (1985)
9. Winograd, T., Fernando, F.: Understanding Computers and Cognition: A New Foundation for Design. Ablex, Norwood (1986)
10. Merrill, D., Twitchell, D.: Instructional Design Theory. Educational Technology Publication Inc., Englewood Cliff (1994)
11. Anderson, J.R.: The Architecture of Cognition. Lawrence Erlbaum Associates Inc., NJ (1996)
12. Collins, A.M., Quillian, R.: Retrieval time from semantic memory. J. Verbal Learn. Verbal Behav. 8, 240–247 (1969)
13. Minsky, M.: A Framework for representing knowledge. In: Winston, P.H. (ed.) The Psychology of Computer Vision. McGraw-Hill, New York (1975)
14. Sowa, J.F.: Conceptual Structures: Information Processing in Mind and Machine. Addison Wesley, Reading (1984)
15. Ramoni, M.F., Stefanelli, M., Magnani, L., Barosi, G.: An epistemological framework for medical knowledge-based systems. IEEE Trans. Syst. Man Cybern. 22, 1–14 (1992)

16. Bloom, B.S.: Taxonomy of Educational Objectives: the Classification of Educational Goals. McKay D, New York (1975)
17. Merrill, M.D.: An introduction to instructional transaction theory. Educ. Technol. **31**, 45–53 (1991)
18. Kyllonen, P.C., Shute, V.J.: A Taxonomy of Learning Skills. Brooks Air Force Base, TX, AFHRL Report No. TP-87–39 (1988)
19. De Jong, T., Ferguson-Hessler, M.G.M.: Types and qualities of knowledge. Educ. Psychol. **31**(2), 105–113 (1996)
20. Schraw, G.J., Olafson, L.: Assessing teachers' epistemological and ontological world views. In: Khine, M.S. (ed.) Knowing, Knowledge and Beliefs, pp. 24–44. Springer, Heidelberg (2008)
21. Peters, J.M., Gray, A.: Epistemological perspectives in research on teaching and learning science. Educ. Res. **5**(1), 1–27 (2006)
22. Brachman, R.: On the epistemological status of semantic networks. In: Findler, N.V. (ed.) Associative Networks: Representation and Use of Knowledge by Computers. Academic Press, New York (1979)
23. Hickman, F.R., Killin, J.L., Land, L., Mulhall, T., Porter, D., Taylor, R.: Analysis for knowledge based systems. In: A Practical Guide to the KADS Methodology. Ellis Horwood, New York (1989)
24. Brown, J.S., Burton, R.R.: SOPHIE: a pragmatic use of artificial intelligence in CAI. In: Proceedings of the 1974 annual ACM conference, vol. 2, pp. 571–579 (1974)
25. Nwana, H.: Intelligent tutoring systems: an overview. Artif. Intell. Rev. **4**, 251–277 (1990)
26. Clancey, W.J.: Overview of Guidon. J. Comput.-Based Instr. **10**(1–2), 8–15 (1982)
27. Corbett, A.T., Koedinger, K.R., Anderson, J.R.: Intelligent tutoring systems. In: Helander, T.K., Landauer, P. (eds.) Handbook of Human-Computer Interaction. Elsevier Science, Amsterdam (1997)
28. Corbett, A.T., Anderson, J.R.: The LISP intelligent tutoring system: research in skill acquisition. In: Larkin, J., Chabay, R., Scheftic, C. (eds.) Computer Assisted Instruction and Intelligent Tutoring Systems: Establishing Communication and Collaboration. Erlbaum, Hillsdale (1992)
29. Beck, J., Mia, S., Haugsjaa, E.: Applications of AI in education. Crossroads Spec. Issue Artif. Intell. **3**(1), 11–15 (1996)
30. Lesgold, A., Lajoie, S., Bunzo, M., Eggan, G.: SHERLOCK: A coached practice environment for an electronics troubleshooting job. In: Larkin, J., Chabay, R. (eds.) Computer assisted instruction and intelligent tutoring systems: shared issues and complementary approaches, pp. 201–238. Lawrence Erlbaum Associates, Hillsdale (1992)
31. Nawaz, M.S., Hassan, M., Shaukat, S.: Impact of knowledge management practices on firm performance: testing the mediation role of innovation in the manufacturing sector of Pakistan. Pak. J. Commer. Soc. Sci. **8**(1), 99–111 (2014)
32. Lytras, M., Pouloudi, A., Poulymenakou, A.: Knowledge management convergence—expanding learning frontiers. J. Knowl. Manage. **6**(1), 40–51 (2002)
33. Wu, I.L., Chen, J.L.: Knowledge management driven firm performance: the roles of business process capabilities and organizational learning. J. Knowl. Manage. **18**(6), 1141–1164 (2014)
34. Nonaka, I.: Dynamic theory of organizational knowledge creation. Organ. Sci. **5**(1), 14–37 (1994)
35. Pathirage, C., Amaratunga, D., Haigh, R.: Tacit knowledge and organizational performance: construction industry perspective. J. Knowl. Manag. **11**(1), 115–126 (2007)
36. O'Dell, C., Grayson, J.: If Only We Knew What We Know. The Free Press, New York (1998)
37. Baumard, P.: Tacit knowledge in professional firms: the teaching of firms in very puzzling situations. J. Knowl. Manag. **6**(2), 135–151 (2002)
38. Borges, R.: Tacit knowledge sharing between IT workers: the role of organizational culture, personality, and social environment. Manag. Res. Rev. **36**(1), 89–108 (2013)
39. Nonaka, I.: The Knowledge Creating Company, pp. 21–46. Harward Business Review on Knowledge Management, USA (1998)

40. Polanyi, M.: The logic of tacit inference. Philosophy **41**(155), 1–18 (1966)
41. Polanyi, M.: Personal Knowledge. Towards a Post Critical Philosophy, Routledge, London (1998)
42. Berente, N.: (2007), http://filer.case.edu/~nxb41/tacit.html
43. Hodgkin, R.: Michael Polanyi—prophet of life, the universe and everything. Times High. Educ. Suppl. 15–21 (1991)
44. Linde, C.: Narrative and social tacit knowledge. J. Knowl. Manag. **5**(2), 160–170 (2001)
45. Chen, L., Mohamed, S.: The strategic importance of tacit knowledge management activities in construction. Constr. Innov. **10**(2), 138–163 (2010)
46. Nonaka, I., Takeuchi, H.: Theory of organisational knowledge creation. In: Theoksessa, H., Takeuchi, H., Nonaka, I. (ed.) Hitutsobashi on Knowledge Management, (2004)
47. Jones, S., Eden, C.: Modelling in marketing: explicating subjective knowledge. Eur. J. Mark. (1981)
48. Cropper, S., Eden, C., Ackermann, F.: Keeping sense of accounts using computer-based cognitive maps. Soc. Sci. Comput. Rev. **8**(3), 345–366 (1990)
49. Huff, A.S.: Mapping Strategic Thought. Wiley (1990)
50. Bougon, M.G.: Uncovering cognitive maps: the self-Q technique. Beyond Method: Strat. Soc. Res. **173487** (1983)
51. Ambrosini, V., Bowman, C.: Tacit knowledge: some suggestions for operationalization. J. Manage. Stud. **38**(6), 811–829 (2001)
52. Rabionet, S.E.: How I learned to design and conduct semi-structured interviews: an ongoing and continuous journey. Qual. Rep. **16**(2), 563–566 (2011)
53. Guarino, N., Welty, C.: Evaluating ontological decisions with OntoClean. Commun. ACM **45**(2), 61–65 (2002)
54. Fortuna, B., Grobelnik, M., Mladenic, D.: OntoGen: semi-automatic ontology editor. In: HCI International 2007, Beijing (2007)
55. Cimiano, P., Völker, J.: Text2Onto–a framework for ontology learning and data-driven change discovery. In: Montoyo, A., Muñoz, R., Métais, E. (eds.) NLDB 2005. LNCS, vol. 3513, pp. 227–238. Springer, Heidelberg (2005)
56. Zouaq, A., Nkambou, R.: Enhancing learning objects with an ontology-based memory. IEEE Trans. Knowl. Data Eng. **21**(6), 881–893 (2009)
57. Zouaq, A., Nkambou, R.: Evaluating the generation of domain ontologies in the knowledge puzzle project. IEEE Trans. Knowl. Data Eng. **21**(11), 1559–1572 (2009)
58. Velardi, P., Navigli, R., Cuchiarelli, A., Neri, F.: Evaluation of ontolearn, a methodology for automatic population of domain ontologies. In: Buitelaar, P., Cimiano, P., Magnini, B. (eds.) Ontology Learning from Text: Methods, Applications and Evaluation. IOS Press, Amsterdam (2005)
59. Buitelaar, P., Cimiano, P., Magnini, B.: Ontology learning from text: an overview. In: Buitelaar, P., Cimiano, P., Magnini, B. (eds.) Ontology Learning from Text: Methods, Evaluation and Applications. Frontiers in Artificial Intelligence and Applications Series, vol. 123. IOS Press, Amsterdam (July 2005)
60. Sabou, M.: Learning web service ontologies: an automatic extraction method and its evaluation. In: Buitelaar, P., Cimiano, P., Magnini, B. (eds.) Ontology Learning from Text: Methods, Evaluation and Applications. IOS Press, Amsterdam (2005)
61. Moldovan, D.I., Girju, R.C.: An interactive tool for the rapid development of knowledge bases. Int. J. Artif. Intell. Tools (IJAIT) **10**(1–2) (2001)
62. Reinberger, M.-L., Spyns, P.: Unsupervised text mining for the learning of DOGMA inspired ontologies. In: Buitelaar, P., Cimiano, P., Magnini, B. (eds.) Ontology Learning from Text: Methods, Applications and Evaluation. Advances in Artificial Intelligence, pp. 29–43. IOS Press, Amsterdam (2005)
63. Salton, G., Buckley, C.: Term-weighting approaches in automatic text retrieval. Inf. Process. Manage. **24**(5), 515–523 (1988)
64. Frantzi, K., Ananiadou, S., Tsuji, J.: The c-value/nc-value method of automatic recognition for multi-word terms. In: Nikolaou, C., Stephanidis, C. (eds.) ECDL 1998. LNCS, vol. 1513, pp. 585–604. Springer, Heidelberg (1998)

65. Harris, Z.: Distributional structure. Word **10**(23), 146–162 (1954)
66. Almuhareb, A., Poesio, M.: Attribute-based and value-based clustering: an evaluation. In: Proceedings of EMNLP, Barcelona (July 2004)
67. Lin, D., Pantel, P.: Induction of semantic classes from natural language text. In: Proceedings Of SIGKDD 2001, San Francisco, CA, pp. 317–322 (2001)
68. Sahlgren, M.: The word-space model: using distributional analysis to represent syntagmatic and paradigmatic relations between words in high-dimensional vector spaces. Ph.D. dissertation, Department of Linguistics, Stockholm University (2006)
69. Lin, D.: Automatic identification of non-compositional phrases. In: Proceeding of ACL 1999, pp. 317–324 (1999)
70. Widdows, D., Dorow, B.: A graph model for unsupervised lexical acquisition. In: 19th International Conference on Computational Linguistics, Taipei, Taiwan, pp. 1093–1099 (2002)
71. Hearst, M., Schutze, H.: Customizing a lexicon to better suit a computational task. In: ACL SIGLEX Workshop, Columbus, Ohio (1993)
72. Hindle, D.: Noun classification from predicate-argument structures. In: Proceeding of ACL 1990, Pittsburg, Pennsylvania, pp. 268–275 (1990)
73. Lin, D.: Automatic retrieval and clustering of similar words. In: Proceeding of COLING-ACL 1998, Montreal, Canada, pp. 768–774 (1998)
74. Cimiano, P.: Ontology learning attributes and relations. In: Ontology Learning and Population from Text, pp. 185–231. Springer, Heidelberg (2006)
75. Fortuna, B., Mladevic, D., Grobelnik, M.: Visualization of text document corpus. In: ACAI 2005 Summer School (2005)
76. Navigli, R., Velardi, P., Cucchiarelli, A., Neri, F.: Quantitative and qualitative evaluation of the OntoLearn ontology learning system. In: Proceeding of the 20th International Conference on Computational Linguistics, Switzerland (2004)
77. Almuhareb, A., Poesio, M.: Finding concept attributes in the web. In: Proceeding of the Corpus Linguistics Conference, Birmingham (July 2005)
78. Guarino, N.: Concepts, attributes and arbitrary relations: some linguistic and ontological criteria for structuring knowledge base. Data Knowl. Eng. **8**, 249–261 (1992)
79. Pustejovsky, J.: The Generative Lexicon. MIT Press, Cambridge (1995)
80. Poesio, M., Almuhareb, A.: Identifying concept attributes using a classifier. In: Proceeding Of the ACL Workshop on Deep Lexical Acquisition, pp. 18–27. Association for Computational Linguistics, Ann Arbor (2005)
81. Woods, W.A.: What's in a link: foundations for semantic networks. In: Bobrow, D.G., Collins, A.M. (eds.) Representation and Understanding: Studies in Cognitive Science, pp. 35–82. Academic Press, New York (1975)
82. Poesio, M., Almuhareb, A.: Extracting concept descriptions from the Web: the importance of attributes and values. In: Buitelaar, P., Cimiano, P. (eds.) Bridging the Gap between Text and Knowledge, pp. 29–44. IOS Press, Amsterdam (2008)
83. Ravi, S., Pasca, M.: Using structured text for large-scale attribute extraction. In: Proceeding Of the 17th ACM Conference on Information and Knowledge Management, CIKM-2008 (2008)
84. Corcho, O., Gómez-Pérez, A.: A Roadmap to ontology specification languages. In: Dieng, R., Corby, O. (eds.) EKAW 2000. LNCS (LNAI), vol. 1937, pp. 80–96. Springer, Heidelberg (2000)
85. Hearst, M.: Automatic acquisition of hyponyms from large text corpora. In: Proceeding of the Fourteenth International Conference on Computational Linguistics, Nantes, pp. 539–545 (1992)
86. Cimiano, P., Volker, J.: Towards large-scale, open-domain and ontology-based named entity classification. In: Proceeding of RANLP 2005, Borovets, Bulgaria, pp. 166–172 (2005b)
87. Snow, R., Jurafsky, D., Ng, A.Y.: Learning syntactic patterns for automatic hypernym discovery. In: Advances in Neural Information Processing Systems (NIPS 2004), Vancouver, British Columbia (2004)

88. Cimiano, P., Staab, S.: Learning by googling. ACM SIGKDD Explor. **6**(2), 24–33 (2004)
89. Maedche, A., Staab, S.: Ontology learning for the semantic web. IEEE Intell. Syst. **16**(2), 72–79 (2001)
90. Etzioni, O., Cafarella, M., Downey, D., Kok, S., Popescu, A.-M., Shaked, T., Soderland, S., Weld, D.S., Yates, A.: Web-scale information extraction in knowitall (preliminary results). In: Proceeding of the 13th World Wide Web Conference, pp. 100–109 (2004)
91. Buitelaar, P., Olejnik, D., Sintek, M.: A protege plug-in for ontology extraction from text based on linguistic analysis. In: Fensel, D., Sycara, K., Mylopoulos, J. (eds.) ISWC 2003. LNCS, vol. 2870. Springer, Heidelberg (2003)
92. Maedche, A., Pekar, V., Staab, S.: Ontology learning part one—on discovering taxonomic relations from the web. In: Web Intelligence, pp. 301–322. Springer, Heidelberg (2002)
93. Bisson, G., Nedellec, C., Canamero, L.: Designing clustering methods for ontology building—the Mo'K workbench. In: Proceeding of the ECAI Ontology Learning Workshop, pp. 13–19 (2000)
94. Caraballo, S.A.: Automatic construction of a hypernym-labeled noun hierarchy from text. In: Proceeding of the 37th Annual Meeting of the Association for Computational Linguistics, pp. 120–126 (1999)
95. Faure, D., Nedellec, C.: A corpus-based conceptual clustering method for verb frames and ontology. In: Velardi, P. (ed.) Proceeding of the LREC Workshop on Adapting lexical and corpus resources to sublanguages and applications, pp. 5–12 (1998)
96. Widdows, D.: Unsupervised methods for developing taxonomies by combining syntactic and statistical information. In: Proceeding of HLT-NAACL, pp. 197–204 (2003)
97. Baker, C.F., Fillmore, C.J., Lowe, J.B.: The berkeley framenet project. In: Proceeding of the COLING-ACL, Montreal, Quebec, Canada (1998)
98. Kipper, K., Dang, H.D., Palmer, M.: Class-Based Construction of a Verb Lexicon. In: Proceeding of AAAI-2000 Seventeenth National Conference on Artificial Intelligence, pp. 691–696 (2000)
99. Cimiano, P., Wenderoth, J.: Automatically learning qualia structures from the web. In: Proceeding of the ACL Workshop on Deep Lexical Acquisition, pp. 28–37 (2005)
100. Claveau, V.: Acquisition automatique de lexiques sémantiques pour la recherche d'information. Thèse de doctorat, Université de Rennes-1, Rennes (2003)
101. Cimiano, P., Wenderoth, J.: Automatic acquisition of ranked qualia structures from the web. In: Proceeding of the 45th Annual Meeting of the Association for Computational Linguistics (ACL), Prague (2007)
102. Craven, M., Di Pasquo, D., Freitag, D., McCallum, A., Mitchell, T.M., Nigam, K., Slattery, S.: Learning to construct knowledge bases from the World Wide Web. Artif. Intell. **1–2**(118), 69–113 (2000)
103. Schlobach, S., Olsthoorn, M., de Rijke, M.: Type checking in open-domain question answering. In: Proceeding of European Conference on Artificial Intelligence, pp. 398–402. IOS Press, Amsterdam (2004)
104. Amardeilh, F., Laublet, P., Minel, J.-L.: Document annotation and ontology population from linguistic extractions. In: Proceeding of the 3rd International Conference on Knowledge Capture, Banff, Alberta, Canada, pp. 161–168 (2005)
105. Tanev, H., Magnini, B.: Weakly supervised approaches for ontology population. In: Proceeding of EACL 2006, Trento, Italy, pp. 3–7 (2006)
106. Fleischman, M.: Automated subcategorization of named entities. In: 39th Annual Meeting of the ACL. In: Student Research Workshop, Toulouse, France (July 2001)
107. Shamsfard, M., Barforoush, A.A.: An Introduction to HASTI: an ontology learning system. In: Proceeding of 6th Conference on Artificial Intelligence and Soft Computing (ASC 2002), Banff, Alberta, Canada (2002)
108. Shamsfard, M., Barforoush, A.A.: The state of the art in ontology learning: a framework for comparison. Knowl. Eng. Rev. **18**(4), 293–316 (2003)

109. Volker, J., Haase, P., Hitzler, P.: Learning expressive ontologies. In: Buitelaar, P., Cimiano, P. (eds.) Ontology Learning and Population: Bridging the Gap between Text and Knowledge. Frontiers in Artificial Intelligence and Applications, vol. 167, pp. 45–69. IOS Press, Amsterdam (2008)

Chapter 3
Pedagogy Modeling

3.1 Introduction to Pedagogy Model

The pedagogy model is the brain of ITS, as it is responsible for making strategic decisions throughout the learning sessions. A strategic decision includes identifying a tutoring strategy, recommending an exclusive course coverage plan, gauging performance parameters, and analyzing the post-tutoring measures (learning gains, learner's emotional state throughout learning, the degree of understandability). It recommends the course structure, tailoring the representation of learning material depending on the information captured in the learner model. This model comprises three adaptation features: the Custom-Tailored Curriculum Sequencing module, Tutoring Strategy recommendation module, and Learner Performance Analyzer module, built into it to facilitate customized tutoring for the learner.

3.2 Preliminary Research on Pedagogy Model in ITS

The pedagogy model is the brain of an ITS/e-learning system, responsible for providing adaptability and personalization features.

3.2.1 Open Education System

Open education gives anyone who wants it, for free, both educational resources (called Open education provides free educational resources (known as Open Educational Resources or OERs) as well as whole courses with a clear, linear structure to anyone who wants them. This is known as "open education." In the second situation, a one-of-a-kind training schedule is devised that specifies what subject to study and when, as well as what exercises should be completed to put what you have learned into practice.

When learning about connectivism, you must engage in activities that are taught in a different manner than the traditional way. Rather than focusing on theory, these types of actions emphasize practice and collaboration among individuals in order to make everyone smarter. Rather than being organized in a linear method, the content in these learning environments is available on the Internet and can be found in a variety of locations. The ability to choose one's own learning route is provided by this method. They are related to connectivist learning theories, which state that knowledge expands when a learner is connected to and needs to share knowledge with others in a learning community [1]. To paraphrase the work of [2], connectivism is a new way of thinking about education since knowledge is disseminated out in a network instead of remaining in the head of a single individual, and learning is the act of recognizing patterns in complex networks of people. When it comes to open learning platforms, they are the most beautiful incarnation of a massive, easily accessible conversation among learners and facilitators. In these scenarios, the learner has always been at the center of the learning process, both as the source of the information and as the one who is being instructed. According to [3], learning occurs when people develop interactions on a social, intellectual, and neurological level. As per connectivism, knowledge is derived from a range of perspectives, and it is defined as the application of network principles to characterize the learning process. In other words, it is straightforward to develop new connections and patterns, and to manage the connections and patterns that are already in place. Connectivism, on the other hand, is a phenomenon of the digital age. When you learn in this manner, you must pay attention to how concepts are formed and connected, which influences how much you know. This type of learning obviously necessitates a great deal of independence on the part of the learner, which can lead to further issues. Some people disagree with the way connectivism operates. Verhagen, for example, believes that because people may choose what they want to learn, they may be creating a learning network that confirms their initial concept rather than challenging and evaluating it. This is how meta-cognitive skills can assist a learner in determining which portions of his or her own learning network are beneficial to him or her and which are detrimental. The learners in these networks are clearly at the center of their learning process. The content taught or delivered does not originate with the lecturer or the school [2].

3.2.2 Massive Online Open Courses

According to connectivist ideology, traditional courses must grow into nonlinear and significantly less organized open learning environments where users can create and share their learning content rather than traditional courses. Teaching and learning in such new open learning settings may be difficult at first due to the unfamiliarity with the system since typical elements may not be noticed, and knowledge may be fragmented due to the ignorance of the system. When you first log in to an online open learning environment, you can feel a little lost. MOOCs are an amazing example of social education in an open system, in which learners voluntarily contextualize and contextualize the knowledge they acquire according to this paradigm. A method of organizing information when faced with a complex and dynamic environment is the "sensemaking" [4] approach, which involves imparting meaning to information gained from the environment through metaphors. There is a close relationship between the act of making sense and the process of learning. Individuals can organize and navigate their activities rather than following pre-established patterns laid forth by instructors. The teacher is still required, but education in an open context is different from traditional online courses and organized curriculum design.

According to [5], the primary reasons for a MOOC's popularity are as follows:

1. The problem-based learning through detailed elucidations
2. The accessibility and also the passion of the instructor
3. Active learning
4. Interaction with peers
5. Beneficial course resources.

MOOCs are classified into two types: cMOOCs and xMOOCs. According to [6], a cMOOC is an online course in which there is no alignment with the course content or the instructor. Instead, the alignment is with the learners and their expertise. On the other hand, massive open online courses (xMOOCs) make considerable use of interactive media and a more behaviorist instructional style that emphasizes individual learning rather than group learning [7]. It has been suggested by Romero [8] that the strategies employed in the most popular MOOCs are akin to those employed in a lecture-based format. Text-based resources, video courses, and forum-based interactions are just a few available options. As a result of exclusively using these tools, there may be little communication between the teacher and the students, and some of the interactions provided by a MOOC may go untapped. The integration of game components, often known as gamification, is a significant feature of many MOOCs. Gamification is the practice of applying game mechanics focused on learning to boost user engagement and retention, and it is a technique that many MOOCs are using. This can be accomplished by using points, achievements, targets, and progress bars that display the user's current position concerning the goals [9]. Although gamification can be applied in a variety of contexts, [10] assert that it is especially important in education:

Game players often exhibit traits such as perseverance, risk-taking, attention to detail, and problem-solving abilities that, in an ideal world, would be displayed in the classroom.

In its most basic definition, gamification is "the use of game design ideas in situations other than those of games" [11]. Using the classification of [12], these gaming mechanisms can be classified as either self-elements (such as points, achievement badges, and levels that are used to compete with oneself and acknowledge self-achievement) or socioeconomic components (such as money) (i.e., leaderboards, where students' achievements are shown in public).

Researchers in psychology developed theoretical frameworks that have proven to be extremely useful for this type of investigation [13, 14]. As an instance, the researcher [15] projected an inclusive four-quadrant model linking learning along with the affective states; subsequently, [16, 17] planned an adaptive control of thought (ACT) cognitive theory that assisted as the hypothetical basis for the extensively used Cognitive Tutor ITS system. In contrast, other researchers and investigators in this field have used ITSs to recognize the various impacts of diverse affective/cognitive states on the learning effects of the individuals involved.

To improve teaching outcomes, pedagogical experts have invested significant work in developing instructional strategies and methods [18–20]. According to [21], constructivist approaches were employed in ITS training, and their pedagogical strategies were examined in detail. Their research discovered that the learning impact was inversely associated with monotony and favorably associated with the flow. They also investigated the relationship between emotional state and the learning process. Many different pedagogical ideas have been explored, including game-based strategy instruction [22], cooperative learning environments [23], and intelligent narrative technologies. These are just a few examples of the many various types of pedagogical concepts that have been investigated [24, 25].

3.3 Path Sequencing of Learning Material in Learning Systems

This section illustrates the preliminary research on path sequencing of learning material in an intelligent tutoring system/e-learning system/and learning management system.

As discussed in Chap. 1, ITS is a rule-based system, and the programmer defines all possible rules that address specific circumstances. These rules indicate that it follows a predefined curriculum and offers remedial actions based on learner activity like human tutors do [26–28].

A depth-first traversal algorithm was used for curriculum sequencing in Knowledge-based Systems [29].

To adjudge prior learner knowledge, the Pretest plays an important role. ITS, developed by Chen [30], offers a personalized learning environment by providing learning material to the learner based on previous knowledge about the course. The pathfinder technique has been used to determine prior knowledge of the learner.

The concept map is another technique that has been used to depict the relationship between the topics and associated sub-topics in the form of nodes and edges, where nodes indicate the issues and boundaries indicate the relation. Nonav and Canas suggest this notion of the concept map. They infer this notion from the theory of Ausubel [31].

ITS, developed by Haoran [32], proposed a solution for determining the learning path of a group of learners. They utilized a profile-based framework for determining an appropriate learning path.

Further advancement in this field uses a data mining technique to mine the meaningful learning path for the learner. This system tracks learner activity during learning and recommends the most suitable learning path. ITS, developed by Hsieh [33], incorporated two methodologies: determining the learning path and recommending the learning path. Initially, the system utilizes the apriori algorithm to generate an initial course coverage plan; they used formal concept analysis to determine the association between the concepts and then adjudged the preferable course coverage plan.

Another [34] proposal uses a fuzzy rule-based technique to determine the association between the list of materials and learner requirements based on web navigation. In the recent development in technologies, some concepts of ontologies, genetic-based algorithms, and artificial neural networks are used to recommend a suitable course coverage plan [35, 36].

ITS, developed by Chen [37], utilizes the nature-inspired algorithm to adjudge the Custom-Tailored learning path. They used two critical parameters for fitness function; one is difficulty level and other associations between the course concepts. Another research work on e-learning systems makes use of the nature-inspired algorithm for determining optimal course coverage plans based on the incorrect response on the Pretest Chen [37] Agbonifo and Obolo [38].

In a recent development, a bio-inspired artificial intelligence, i.e., Ant Colony Optimization (ACO), is utilized for determining the course coverage plan. The ant-based system [39] initially utilized the Traveling Salesman Problem (TSP). This mapping helps to determine the optimized learning path. This learning path is represented with the help of graphs with weighted edges that indicates ant pheromone (students), i.e., released pheromones along their path.

In [40], the author uses ACO techniques to recommend an adaptive learning path by considering the learner's Learning Style. Bonabeau et al. [41] utilizes self-organizing techniques to recommend optimal course coverage plans to the learner. Similar practices can be applied through a probabilistic approach. Nodes indicate the pedagogy items, edges indicate the hypertext links (preferred probabilities), and learners act as ants who have to traverse all the nodes [26, 42, 43].

ITS developed by Jamon et al. utilizes an "ant-hill" nature-inspired algorithm based on learner success/failure ratio for validating an item/topics/concepts ant lay pheromone. Thus, the pheromone helps determine the optimized course coverage plan for learners [44]. A style-based Ant Colony system (SACS) utilized an advanced Ant Colony algorithm. They design the user model by determining a suitable course coverage plan for a group of learners using a graph-based path structure [45, 46].

ITS developed by Sengupta et al. experiments with Ant Colony Optimization to attain the personalization feature in the learning path recommendation. They utilized frequent graph patterns to determine the correlation between the concepts [47].

ITS developed by Kardan experiments with a two-phase learning path algorithm. In the first phase, they adjudge the knowledge level of learners based on the performance in the Pretest. In the second phase, they used a meta-heuristic algorithm to determine and recommend a suitable learning path for the learner. [48]

ITS developed by agbonifo et al. experiments with the learner model. Their developed ITS determine the learner's Learning Style using *the "Honey and Mumford"* Learning Style model. To recommend a suitable learning path for learners, they utilize the Neuro-Fuzzy technique. Here, researcher did not consider the difficulty level of the learning material [49, 50].

3.4 Impact of Emotion Capturing in Learning System

ITSs are a generation of a computer-based software system; the purpose is to improve and support learning in a particular domain. Here, an adaptive learning environment indicates the integration of cognitive intelligence into the traditional CAI. It emulates human intelligence, offers the benefits of personal teaching, and also provides a personalized and adaptive learning environment.

For the era, the notion of ITS has been grounded on the principles of constructivism and cognitivism. They are primarily focusing on learners' cognitive processes. Recently, researchers have switched their attention from learner's cognitive processes to learner's emotion-enabled cognitive processes. This switching is due to the researcher emphasizing the correlation between learning and emotions. Previous research shows that emotions play an ample role in the learning process as they are equally responsible for affecting learners' learning and motivation abilities. Previous research studies deduce that learners perceive negative and positive sentiments throughout the learning process. Thus, it indicates that more emphasis is to be given to learners' emotion-enabled cognitive processes in the development process of the learning system.

3.4.1 Emotion Recognition in Learning System

Murthy and Jadon looked at six types of emotions: Sad, Happy, Surprise, Disgust, Normal, and Ambiguous. Eigenfaces are used to tell how someone is feeling. They wanted to use the dimensionality-reduction method (Principal Component Analysis) on a bigger dataset [50]. They were able to get an 83% success rate using this method. Using PCA makes this method more expensive because it takes a lot of time to figure out the correlation matrix. This is especially true when a lot of data is used to train. Lien and other people looked at two methods: SVD (Singular Value Decomposition) and explicit matching [51, 52]. They started by changing the images into suitable transitional expression matrices; they did a direct match. Because a direct matching procedure does not give you much precision when you figure out correlation coefficients, these two methods do not work well for many things. As a result, facial image conversion would make facial images that were not symmetrical [53]. At the same time, 79, Arumugam combines FLD, Fisher's Linear Discriminant, and SVD, Singular Value Decomposition. The Radial Basis Function is used as a classifier. They primarily focused on three types of emotions: disgust, happiness, and anger. The main problem with this strategy is that it does not work very well. The speed and power of many processors cannot keep up with naive SVD [54, 55].

A computer program called the Itspoke ITS talks to a learner through a long physics qualitative inquiry. It tells the learner where they are not clear. They use sound and prosodic elements from the learner's speech to figure out their emotional state, which can be positive, neutral, or negative. They were able to get an accuracy rate of 80.53 percent using this method. It also talks about more complex emotions, like perplexity, boredom, and frustration. AutoTutor [56] talks about these more detailed emotions. They could tell how someone was feeling by how they moved, spoke, and facial features looked [57].

They use an animated pedagogical agent with animated facial expressions, sounds, and speech to get correct answers from a tutoring system that can help people learn. There are five emotions Woolf has talked about: self-confidence, impatience, boredom, and exhaustion. He used a lot of heuristic rules to respond to the learner's thought process (changing voice and gesture, empathetic response, graphs and hints, text messaging). He looked at the level of involvement concerning how it affected learning and behavior. Mao and Li [58] and Li [59] came up with "ALICE," an Emotion-Sensitive ITS. Alice uses an emotion agent, which can read text, speech, and facial expressions to determine how the learner feels. They worked with human teachers to think about all possible situations and develop rules. As a result of the many tutoring sessions, ALICE acts the most like a real person.

Tian [60] came up with a system that used active listening and emotional computing. This framework can determine how people feel by looking at text interactions like typed words, group conversations, chat rooms, and question-and-answer. They use case-based reasoning to come up with an effective text-based answer. Strain and D'Mello [61] did an analytical study on a learner's psychological state while performing any task. Systems initially consider the negative emotion of learners toward learning sessions and begin learning sessions accordingly, and then answer the questions about what they had learned. The results show that cognitive reappraisal as an emotion regulation strategy leads to more positive activating emotions and better reading comprehension.

As mentioned earlier, minimal work concerning learner path sequencing is reported so far in ITS—a couple of instances discussed below.

ELM-ART ITS [26] provides path sequencing and adaptive hypermedia, enabling the learner to navigate this material. ELM-ART provides navigation support by utilizing two hypermedia techniques, i.e., adaptive annotation (focus on visual artifacts and their representation) and adaptive sorting (determine the similarity between concepts and offer more relevant concepts).

3.5 Summary

This chapter explored the work in Intelligent Tutoring Systems (ITS)/E-learning systems/Hypermedia systems for the pedagogy domain. The pedagogy model that is considered the heart of the ITS has been explored, emphasizing learning path sequencing techniques used to provide adaptivity and personalization in the ITS/e-learning system/ LMS. Toward the concluding part of this chapter, the importance of emotion recognition in the ITS/e-learning system/ LMS is discussed.

References

1. Kop, R., Hill, A.: Connectivism: learning theory of the future or vestige of the past? [online] http://www.irrodl.org/index.php/irrodl/article/view/523/1103 (2008). Accessed 05 May 2015
2. Siemens, G.: Connectivism: learning theory or pastime of the self-amused? Elearnspace blog [online] http://www.elearnspace.org/Articles/connectivism_self-amused.htm (2006). Accessed 08 Oct 2014
3. Siemens, G.: What is the unique idea in connectivism [online] http://www.connectivism.ca/?p=116 (2008). Accessed 08 Oct 2014
4. Siemens, G.: What is connectivism? [online] http://docs.google.com/Doc?id=anw8wkk6fjc_14gpbqc2dt (2009). Accessed 17 Mar 2015
5. Hew, K.F.: Promoting engagement in online courses: what strategies can we learn from three highly rated MOOCs. Br. J. Edu. Technol. **4**(2), 320–341 (2015)
6. Rodriguez, C.O.: MOOCs and the AI-Stanford like courses: two successful and distinct course formats for massive open online courses. Eur. J. Open. Distance and E-Learning [online] http://www.eurodl.org/materials/contrib/2012/Rodriguez.pdf (2012). Accessed 05 May 2016
7. Conole, G.: A new classification schema for MOOCs. Int. J. Innov. Qual. Learn. **2**(3), 65–77 (2014)
8. Romero, M.: Game based learning MOOC. Promoting entrepreneurship education. Elearning Pap. Spec. Ed. MOOCs Beyond **33**, 1–5 (2013)
9. Parry, M.: 5 Ways that edX could change education. Chronicle High. Educ. [online] http://chronicle.com/article/5-Ways-That-edX-Could-Change/134672 (2012). Accessed 18 Mar 2015
10. Klopfer, E., Osterweil, S., Salen, K.: Moving learning games forward. Obstacules, opportunities and openness. The Education Arcade, Massachussets Institute of Technology [online] http://education.mit.edu/wpcontent/uploads/2015/01/MovingLearningGamesForward_EdArcade.pdf (2009). Accessed 02 Apr 2015
11. Deterding, S., Dixon, D., Khaled, R., Nacke, L.: From game design elements to gamefulness: defining gamification. In: Proceedings of the 15th International Academic MindTrek Conference: Envisioning Future Media Environments, pp. 9–15. (2011)
12. Huang, W.H.Y., Soman, D.: Gamification of education. Res. Rep. Ser. Behav. Econ Action. (online) (online) http://inside.rotman.utoronto.ca/-behaviouraleconomicsinaction/files/2013/09/-GuideGamificationEducationDec2013.pdf (2013). Accessed 30 Jan 2017
13. Arroyo, I., Cooper, D.G., Burleson, W., Woolf, B.P., Muldner, K., Christopherson, R.: Emotion Sensors go to School. IOS Press (2009)
14. D'Mello, S., Graesser, A., Picard, R.W.: Toward an afectsensitive autotutor. IEEE Intell. Syst. **22**(4), 53–61 (2007)
15. Kort, B., Reilly, R., Picard, R.W.: An affective model of interplay between emotions and learning: reengineering educational pedagogy-building a learning companion. In: Proceedings IEEE International Conference on Advanced Learning Technologies. IEEE (2002)
16. Anderson, J.R.: Cognitive Psychology and its Implications. Freeman, San Francisco (1980)

17. Anderson, J.R.: The Architecture of Cognition. Harvard University Press, Cambridge, Mass (1983)
18. Kolodner, J.: Facilitating the learning of design practices: Lessons learned from an inquiry into science education. J. Ind. Teach. Educ. **39**(3), 9–40 (2002)
19. Luckin, R., Holmes, W., Grifths, M., Forcier, L. B.: Intelligence unleashed. An argument for AI in education. Pearson (2016)
20. VanLehn, K.: The behavior of tutoring systems. Int. J. Artif. Intell. Educ. **16**, 227–265 (2006)
21. Graesser, A.C., Chipman, P., Haynes, B.C., Olney, A.: Autotutor: an intelligent tutoring system with mixed-initiative dialogue. IEEE Trans. Educ. **48**(4), 612–618 (2005)
22. Lester, J.C., Ha, E.Y., Lee, S.Y., Mott, B.W., Rowe, J.P., Sabourin, J.: Serious games get smart: intelligent game-based learning environments. AI Mag. **34**(4), 31–45 (2013)
23. Santhanam, R., Liu, D., Shen, W.C.: Research note gamifcation of technology-mediated training: not all competitions are the same. Inf. Syst. Res. **27**(2), 453–465 (2016)
24. McCoy, J., Treanor, M., Samuel, B., Wardrip-Fruin, N., Mateas, M.: Comme Il Faut: a system for authoring playable social models. In: Proceedings of the Seventh International Conference on Artificial Intelligence and Interactive Digital Entertainment, pp. 158–163. AAAI Press, Palo Alto, CA (2011)
25. Yu, H., Riedl, M.O.: A Sequential recommendation approach for interactive personalized story generation. In: Proceedings of the Eleventh International Conference on Autonomous Agents and Multiagent Systems, pp. 71–78. International Foundsation for Autonomous Agents and Multiagent Systems, Richland, SC (2012)
26. Brusilovsky, P., Schwarz, E., Weber, G.: ELMART: An intelligent Tutoring System on World Wide Web. In: Frasson, C., Gauthier, G., Lesgold, A. (eds.) Third International Conference on Intelligent Tutoring Systems, ITS-96, vol. 1086, pp. 261–269. Springer, Berlin (1996)
27. Skinner, B.F.: Teaching machines. Science **128**
28. Wu, D.: Active acquisition of user models: implications for decision-theoretic dialog planning and plan recognition. In: User Modeling and User-Adapted Interaction, pp. 149–172 (1991)
29. Baldoni, M., Baroglio, C., Patti, V.: Structureless, intention-guided web sites: planning based adaptation. In: Proceedings 1st International Conference on Universal Access in Human-Computer interaction, vol. 3, pp. 237–241. (2001)
30. Chen, L.H.: Enhancement of student learning performance using personalized diagnosis and remedial learning system. Comput. Educ. **56**, 289–299 (2011)
31. Ausubel, D.P.: The psychology of meaningful verbal learning. (1963)
32. Haoran, X., Di, Z., Fu, L.W., Tak-Lam, W., Yanghui, R., Simon, H.W.: Discover learning path for group users: a profile-based approach. Neurocomputing (2017)
33. Hsieh, T.-C.: A mining based approach on discovering courses pattern for constructing suitable learning path. Expert Syst. Appl. **37**, 4156–4167 (2010)
34. Li, Y., Huang, R.: Dynamic composition of curriculum for personalized e-learning. Knowledge Science & Engineering Institute, Beijing Normal University (2006)
35. Karampiperis, P., Sampson, D.: Adaptive instructional planning using ontologies. In: Proceedings of the IEEE International Conference on Advanced Learning Technologies (ICALT'04). (2004)
36. Dron, J.: Achieving self-organisation in network-based learning environments. Phd thesis. University of Brighton (2002)
37. Chen, C.M.: Intelligent web-based learning system with personalized learning path guidance. Comput. Educ. **51**(2), 787–814 (2008)
38. Agbonifo, O.C., Obolo, O.A.: Genetic algorithm-based curriculum sequencing model for personalised e-learning system. (2018)
39. Seridi, H., Sari, T., Sellami, M.: Adaptive instructional planning in intelligent learning systems. In: Proceedings of the IEEE International Conference on Advanced Learning Technologies (ICALT'06). (2006)
40. Hong, C.-H., Chen, C.-M., Chang, M.-H., Chen, C.-H.: Intelligent web-based tutoring system with personalized learning path guidance. In: Proceedings of the IEEE International Conference on Advanced Learning Technologies (ICALT'07). (2007)

41. Bonabeau, E., Dorigo, M., Theraulaz, G.: Swarm intelligence: from natural to artificial systems. Oxford University Press (1999)
42. Dahbi, A., Elkamoun, N., Berraissoul, A.: Adaptation and optimisation of pedagogical paths by ant's algorithm. In; IEEE Information and Communication Technology (ICTTA'06). (2006)
43. van den Berg, B., van Es, R., Tattersall, C., Janssen, J., Manderveld, J., Brouns, F., Kurves, H., Koper, R.: Swarm-based sequencing recommendations in e-learning. In: Proceedings of the 2005 5th International Conference on Intelligent Systems Design and Applications (ISDA'05). (2005)
44. Jamont, Y., Valigiani, G., Biojout, R., Lutton, E.: Experimenting with a real-size man-hill to optimize pedagogical path. In: SAC'05 Proceedings, USA (2005)
45. Semet, Y., Yamont, Y., Biojout, R., Luton, E.: Ant Colony Optimization for e-learning: observing the emergence of pedagogic suggestions. In: IEEE Swarm Intelligence Symposium. (2003)
46. Wang,T.-I., et.al.: Using a style based ant colony system for adaptive learning. Expert Syst. Appl. https://doi.org/10.1016/j.eswa.2007.04.014
47. Ahmed, A.K, Molood, A.E, Maryam, B.I.: A new personalized learning path generation method: ACO-MAP. Indian J. Sci. Res. 5(1), (2014)
48. Agbonifo, O.C., Adewale, O., Alese, B.K.: Design of a neurofuzzy-based model for active and collaborative online learning. Unpublished Ph.D. thesis, Federal University of Technology Akure, pp. 7–8 (2012).
49. Adesuyi, A.T., Adewale, O.S., Thompson, A.F.: Ontology-Based personalisation system for e-learning. Int. J. Comput. Sci. Eng. Technol. 1(1), 1–11 (2014)
50. Murthy, G.R.S., Jadon, R.S.: Effectiveness of eigen spaces for facial expressions recognition. Int. J. Comput. Theory Eng. 1(5), 638–642 (2009)
51. Becker, B.C., Ortiz, E.G., Flori, C.: Evaluation of face R recognition techniques for application to facebook 5000 forbes Avve university of Centr. In: 8th IEEE International Conference on Automatic Face and Gesture Recognition, pp. 1–6. (2008)
52. Lien, C., Chang, Y., Tien, C.: A fast facial expression recognition method at low-resolution images. In: International Conference on Intelligent Information Hiding and Multimedia, pp. 419–422. (2006)
53. Arumugam, D.S.P.: Emotion classification using facial expression. Int. J. Adv. Comput. Sci. Appl. 2(7), (2011)
54. Lu, H., Plataniotis, K.N.K., Venetsanopoulos, A.N.: MPCA: mulitilinear proncipal component analysis of tensor objects. IEEE Trans. Neural Netw. 19(1), 18–39 (2008). A publication of the IEEE Neural Networks Council
55. Litman, D., Forbes, K.: Recognizing emotions from student speech in tutoring dialogues. In: Proceedings of the ASRU, 2003.
56. D'Mello, S., Jackson, T., Craig, S., Morgan, B., Chipman, P., White, H., Person, N., Kort, B., Kaliouby, R., Picard, R.W., Graesser, A.: AutoTutor detects and responds to learners affective and cognitive states. In: Workshop on Emotional and Cognitive Issues at the International Conference of Intelligent Tutoring Systems, 2008.
57. Woolf, B., Burleson, W., Arroyo, I., Dragon, T., Cooper, D., Picard, R.: Affect-aware tutors: recognising and responding to student affect. Int. J. Learn. Technol. 4(3/4), 129 (2009)
58. Mao, X., Li, Z.: Implementing emotion-based user-aware e-learning. In: Proceedings of the 27th international conference extended abstracts on Human factors in computing systems— CHI EA '09 (pp. 3787–3792). ACM Press, New York, NY (2009)
59. Mao, X., Li, Z.: Agent based affective tutoring systems: a pilot study. Comput. Educ. 55(1), 202–208 (2010)
60. Tian, F., Gao, P., Li, L., Zhang, W., Liang, H., Qian, Y., Zhao, R.: Recognizing and regulating e-learners' emotions based on interactive Chinese texts in e-learning systems. Knowl.-Based Syst. 55, 148–164 (2014)
61. Strain, A., D'Mello, S.: Emotion regulation during learning. In: Biswas, G., Bull, S., Kay, J., Mitrovic, A. (eds.) Artificial Intelligence in Education SE-103 (vol. 6738, pp. 566–568). Springer, Berlin Heidelberg (2011). https://doi.org/10.1007/978-3-642-21869-9_103

Chapter 4
Building SeisTutor Intelligent Tutoring System for Experimental Learning Domain

4.1 Introduction

Domain model is the heart of an Intelligent Tutoring System (ITS) because the domain model contains the expert knowledge, which is gained through experience and makes available for the novice learner when requested. This model is highly focused on the "What to Teach" issue (see Fig. 4.1). Domain model comprises "what" is to be taught. The learning material forms content of course (SDI). Knowledge Engineer gathers domain knowledge from the domain experts, realigns, and sequences the gathered knowledge under the supervision of the subject domain experts. The aim of the domain model is to keep the subject domain and the learning material as content in the course and deliver it to the learner. This model organizes the subject topic, sub-topics, and their association with other topics.

The organizing of knowledge is in a style and level as per the convenience of the learner, to provide better learning gain.

The presented work considers seismic data interpretation (SDI) as a subject domain. This subject domain is one of the knowledge domains, which is undocumented and highly experiential. Since the knowledge is gained through experience, without any documented sources available, it is qualified as a Tacit Domain of Knowledge.

4.2 Seismic Data Interpretation: As Experiential Learning Domain

Seismic data interpretation (SDI) is the sub-field of Geophysics under the field of Petroleum Exploration. The Petroleum Exploration process uses seismic data to understand and delineate the sub-surface geology. The final result of the exploration process is further processed and is available for analysis in the form of a SEG-Y map of seismic images. Till date, there are human experts, i.e., Seismologists that

N. Singh et al., *Cognitive Tutor*, Advanced Technologies and Societal Change,
https://doi.org/10.1007/978-981-19-5197-8_4

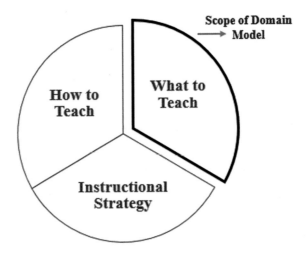

Fig. 4.1 Responsibility of domain model

analyze SEG-Y maps and discover their understandings. During analysis, experts are trying to understand what kind of sub-surface geology exists. This process is termed as interpretation. There are no documented thumb rules for interpretation. The interpretation knowledge available is "implicit" within the experts and exists as their mental database. It has been gained by them, over years of experience, and the time spent in the fields. This knowledge is termed as "Tacit Knowledge." The lack of documentation causes the uncertainty to exist, in interpretation. There may be a possibility that the same SEG-Y map may be interpreted differently by different seismologists, because this knowledge, to a great extent, is highly individualistic and varies from expert to expert.

A novice seismologist joining an organization engaged in Petroleum Exploration processes holds very minimal interpretation powers and needs time and practice to hone satisfactory skills. But this process is time-consuming and expensive intensive. The organizations, incur expenditure on long training cycles, to train them, and let them practice their skills over a period of time, so that they can deliver reasonable interpretation. The major bottleneck is the nature of knowledge, which due to its individualistic and experiential characteristics, incurs long training periods. The focus of this research work is capturing this experiential knowledge, which is present in the tacit/implicit form, converting this knowledge into an explicit form and making it tutor-able. The main reason to explicate this knowledge is to reduce the overall training cycle. Figure 4.2 describes the distinct features of Tacit and Explicit Knowledge.

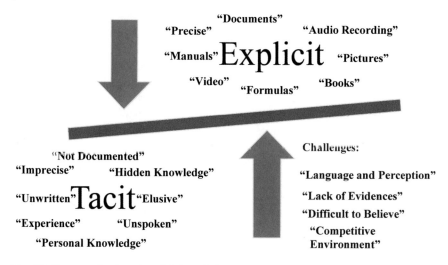

Fig. 4.2 Features of tacit and explicit knowledge

4.3 Development of Adaptive Domain Model

This section describes the development of an adaptive domain model or an adaptive knowledge base. The developed domain model comprises twelve pedagogy styles. Each of the pedagogy styles is distinct to each other. Based on learner preferences, most optimal pedagogy styles are chosen. This feature of SeisTutor makes the domain model an adaptive knowledge base, because it enables the SeisTutor to adjudge the learners' preferences and provide the learning material accordingly [1].

For developing the adaptive domain model, the presented work has been categorized into two phases (see Fig. 4.3).

The objective of the first phase is to acquire tacit knowledge, characterize it, and convert it into an explicit form. In subsequent phases, the explicated knowledge is transformed, into a tutor-able form. For this purpose, the content is shaped as a course with a structured course plan (also referred as, course coverage plan).

Fig. 4.3 Conceptual diagram of the development of the adaptive domain model

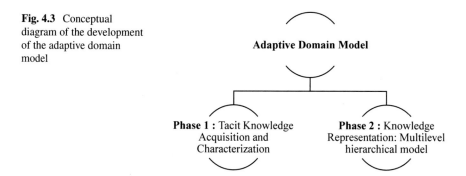

The content coverage organized into units and sub-units, spread over the coverage time, is organized. Various acquisition techniques, i.e., causal map and one-to-one interviews/semi-structured methods, have been used for each of the above steps.

The objective of the second phase is to align the learning material as per the curriculum and make it tutor-able. The packaging of learning material is known as a knowledge capsule. In the present work, the SeisTutor comprises twelve knowledge capsules. Every knowledge capsule holds same content, but their content representation, usage of rich media techniques, and the content elucidation level vary. In several ways, there is variation from capsule to capsule. Therefore, as a result, twelve adaptive knowledge capsules have been developed to offer to the learners, as per the adjudged requirement.

Figure 4.3 describes the conceptual workflow that has been used to develop an adaptive domain model. Subsequent sections describe the techniques and methodologies used in Phase 1 and Phase 2.

4.3.1 Phase 1: Tacit Knowledge Acquisition and Characterization

This section describes the steps followed to discover and gather tacit knowledge from the experts through causal and semi-structured methods. Figure 4.4 shows the process involved to solicit tacit knowledge from the expert and then make it tutor-able. In knowledge gathering step, knowledge from basic to advanced levels, on SDI, is gathered from experts. In knowledge characterization step, the nature of the knowledge is studied, to clarify and elaborate. The related topics and sub-topics are combined, to make a complete unit/package. In knowledge sequencing, the characterized units are re-aligned to form a complete course coverage plan. In validation step, the designed curriculum is verified from the domain experts for their approval. Then, the learning material is organized as individual knowledge capsules as per the validated course coverage plan.

For gathering knowledge on the SDI domain, formal permission was sought from public sector companies who have been involved in seismic data acquisition and interpretation tasks. A total of 09 meetings were held with Deputy General Manager (DGM) and their team. A well-structured questionnaire (causal map construction technique) and one-to-one interviews (semi-structured method) have been used as instruments for domain knowledge collection and conversion.

4.3.1.1 Conversion of Tacit to Explicit: Findings

This section describes the experience during gathering tacit knowledge (of seismic data interpretation) and their transformation into an explicit form.

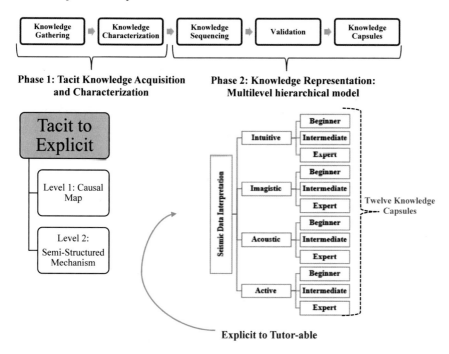

Fig. 4.4 Phases involved to solicit tacit knowledge from experts and make it tutor-able

A formal **note-sheet** presenting the general motivation behind the activity is conveyed to the concerned authority. For acquiring tacit knowledge from experts, two independent groups of 2 individual each is created and the task of requesting tacit knowledge from 5 geologists/seismologists was performed. Out of the group of 5 seismologists, 2 were expert profiles (engaged in interpretation skill for more than 12 and 13 years, respectively), 2 were middle-level expert profiles (who had been doing interpreting activities since 3 and 4 years individually), and the last individual was a novice student (recently in-cumbered into the group to learn and to practice hands-on activity on, interpretation skill).

The aim of the formation of a group with varied proficiency of interpretation skill is purposeful, to accumulate all viewpoints and to have enough ways for iterative data gathering and validation. Over a progression of meetings, the process of obtaining and requesting tacit knowledge from geologists/seismologists is performed. Primarily, the causal map is used, in which a series of the questionnaire were made and put forward in front of domain experts by both the groups and accumulate sufficient information which further helps in the construction of the causal map. The first group did their scrutinizing, probing, and causal map development activity. In the causal map, nodes indicate the questions or topics and links indicate the feature of the nodes (normal, important, and prediction task). This procedure has taken the time of three weeks by group 1 and the other group took around four weeks of time. To accumulate adequate information, questions ranging from techniques used

Table 4.1 Partial list of a questionnaire for causal map construction

S. No.	Questions
i	What does the seismic snap indicate?
ii	Where might you find additional seismic information to affirm the shape and size of the structural trap that you have mapped?
iii	Kindly explain the scenario where the horizon and faults incorrectly interpreted?
iv	Illustrates the steps involved in the identification of faults and their types, through velocity correlation?
v	What are the challenges faced during analysis and interpretation?
vi	Several stories are revealing that error encounters during an acquisition? Please discuss

for seismic data acquisition to analysis to interpretation were developed. Table 4.1 indicates the partial list of questionnaires unfolding the seismic data interpretation process.

4.3.1.2 A Partial List of a Questionnaire for Causal Map Construction

The causal map built by each group is closely examined together to create a reasonably appropriate material representing the conversion of tacit knowledge into explicit knowledge. This piece of work is identified as Level 1.

Figure 4.5 shows the causal map in which the subject theme is ranging from seismic data acquisition and processing to analysis and interpretation. The scope of the presented work is to capture the tacit knowledge of seismic data interpretation, whose terms are numbered, from 19 to 23.

Figure 4.6 shows the graphical representation of the interpretation task. For simplicity, faults and horizons are referenced here. Figure 4.6 reveals that the whole process of interpretation is iterative; with each section repeatedly examines. After accumulating all the translated seismic section sub-surface, the topographical map is created.

The built causal map has been further detailed. The exhaustive causal map (step numbered) has been shown in Fig. 4.7. Figure 4.7 is the continuation of Fig. 4.5 from step number 19, which is numbered from 1 to 9. The depiction of steps recorded beneath:

The semi-structured mechanism is utilized in Level 2 and proceeded with a similar group and same geologist/seismologist groups, as in Level 1. The geologists/seismologists were scrutinized and requested, to share their skills, discoveries, and stories, jot down what worked and what did not and their portrayals were recorded. The two groups drew out their outcomes from the recorded experience stories and narrations. This activity was finished in 5 weeks by the two groups.

The outcomes were cumulated, and a proper record of each topic and sub-topics of interest was created. This notion prompted the development of a knowledge capsule. The significant commitment of the present work is to explicate the tacit

1.   Seismic Acquisition
2.   Onshore
3.   Offshore
4.   Thumper Truck
5.   Geophone
6.   Air gun
7.   Hydrophone
8.   Analog Recording
9.   Analog to digital conversion
10. Wiggle Trace
11. CDP Gather
12. Stacking
13. Normal Move-out correction
14. Processed seismic section
15. Common Depth points
16. Floating Datum
17. TWT (Two Way Time)
18. Time versus Depth Conversion
19. Seismic Map
20. Check-line scale and orientation
21. Top-down approach for clarity of understanding
22. Determine primary reflectors and geometrics
23. Prediction of Hydrocarbon amassing

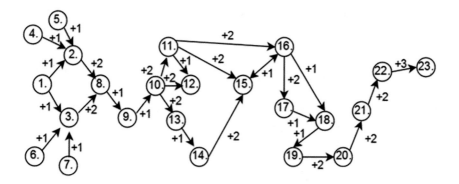

Note   +1 :- Normal Task , +2 :- Important Task , +3 :- Prediction Related Task

Fig. 4.5 Causal map for processes involved in "seismic interpretation domain"

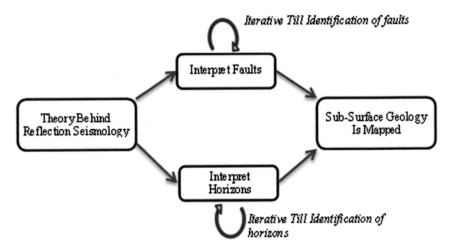

Fig. 4.6 Schematic diagram of seismic interpretation task

knowledge in the form of knowledge capsules, enabling its future scope in phase 2. Figure 4.8 demonstrates the diagrammatic representation of a semi-structured mechanism, where the nodes 1, 14, 19, and 23 are shown, that holds the information that should be transformed from implicit to explicit form. These nodes are further probed. Let us consider the probing of node 1, i.e., 1.A, 1.B, 1.C. Suppose if node one is focused on a specific topic, further detailing on that topic are 2.A, 2.B, 2.C, that might be the supporting sub-topics to briefly describe the concepts of node one or topic one (where 1, 2, 3, and 4 indicate the level of probing while A, B, and C represent the supporting sub-topics of particular node). The primary motivation for adopting this mechanism is to get a clear understanding of the topics. Each detailing (probing) attempted has its purpose; the purpose of node 1 is to comprehend the different strategies to accomplish.

A similar procedure is pursued for the remaining activities (topics/subtopics) 14, 19, till 23. This way, the probing can proceed with further revealed hidden details and eliciting experiences. Figure 4.8 shows the probing of 4 topics up to 4 levels of segregation. The course coverage plan was designed as instructed by the domain experts. Furthermore, the captured domain knowledge has been transformed into knowledge capsules. The development of knowledge capsules puts one level ahead from the explicit representation of the tacit knowledge to tutor-able form. This work is accomplished in phase 2.

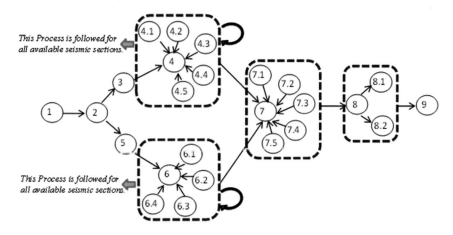

1. Seismic Map
2. Interpretation initiates
3. Structural
4. Structural analysis
5. Stratigraphic Interpretation
6. Stratigraphic analysis
4.1. Faults and folds
4.2 Salt
4.3 Shale Diapers
4.4 Structural Trends
4.5 Structural Features
6.1 Unconformities
6.2 Stratal Packages
6.3 Environments/ Facies/ Lithologies
6.4 Ages
7. Identify Prospect elements
7.1 Source of the geological feature
7.2 Migration
7.3 Reservoir
7.4 Trap
7.5 Seal
8. Assess the Highest Potential Prospects
8.1 How much oil/gas do we except?
8.2 How certain are they?
9. Economical analysis

Fig. 4.7 Detailed causal map process involved to interpret the seismic snap or map

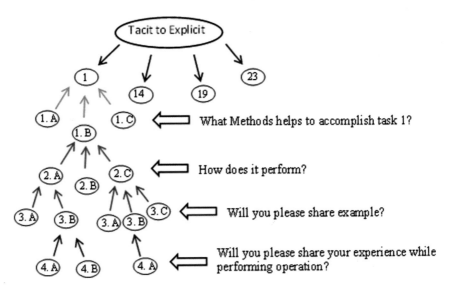

Fig. 4.8 Semi-structured mechanism

4.3.2 Phase 2: Knowledge Representation: Multilevel Hierarchical Model

This section describes the conversion of explicit SDI domain into a tutor-able form in the form of twelve knowledge capsules. Thus, to accomplish the objective, the course manager and the knowledge repository sub-modules have been developed that combine to form a knowledge base module. The purpose of the course manager is to align the subject topics and sub-topics based on the association between them. The purpose of the repository is to warehouse the learning material and assessment material. Learning material and assessment material are described by meta-description that aid in maintaining, locating, representation, and reusability of subject knowledge in the knowledge base [1].

4.3.2.1 Course Manager

The course manager is the graphic representation, of course material. It utilizes data structure techniques for its illustration, i.e., Concept Dependency graph and Concept Tree (see Figs. 4.9, 4.10, and 4.11).

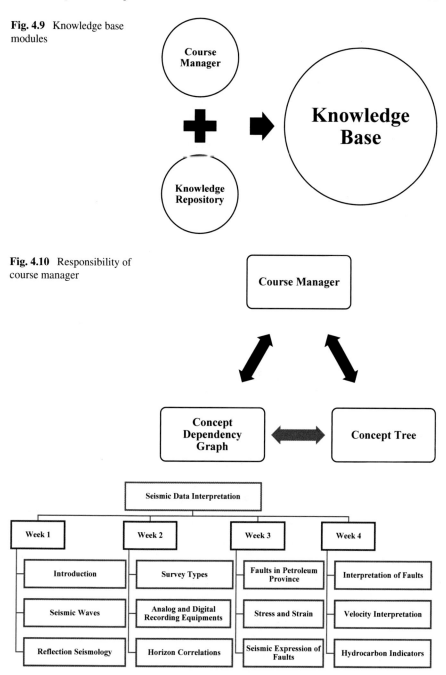

Fig. 4.9 Knowledge base modules

Fig. 4.10 Responsibility of course manager

Fig. 4.11 Course tree representation of subject domain "seismic data interpretation"

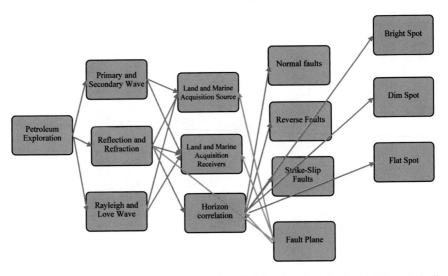

Fig. 4.12 Course dependency graph representation of subject domain "seismic data interpretation"

Concept Tree

The concept tree possesses a hierarchical tree-type data structure, where the subject name resides at the root node and topics and sub-topics reside on a leaf node (see Fig. 4.11).

Concept Dependency Graph

The concept dependency graph is the graphical representation of the association between subject topics and sub-topics (see Fig. 4.12). Every domain knowledge warehoused in the knowledge base is the pair representation of concept dependency graph and concept tree which work together and aid in maintaining, locating, representing, and reusability of subject knowledge.

4.3.2.2 Knowledge Repository

Knowledge repository comprises learning materials and assessment materials. As aforementioned, every learning topic and sub-topics provides meta-description which speeds up the accessing of the learning material from the knowledge pool. The critical issues in the teaching system are overwhelmed by ontologies [2]. The interim web offers a productive, interactive mode of learning by utilizing audio, text, images, pedagogical agents, and animation, which persuades the active interaction between the learner and computer-aided system [3].

Furthermore, this active interaction enriches the persuasive communication and improves problem-solving skills and overall learning gain. The ontology-based representation of the subject domain is demonstrated in Fig. 4.13. This representation aids in enriching the recommendation of learning materials. The amalgamation of learning material with this kind of representation aids the SeisTutor to pacify the learner preferences (mode of learning, the difficulty level). In this presented work, a three-level knowledge base frame is used, that is demonstrated in Fig. 4.14.

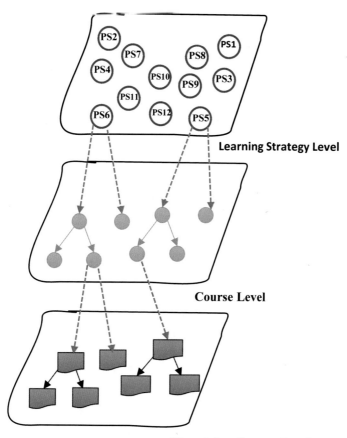

Fig. 4.13 Knowledge base frame

Fig. 4.14 Domain
pedagogical structure in
SeisTutor (learning style and
learner profile)

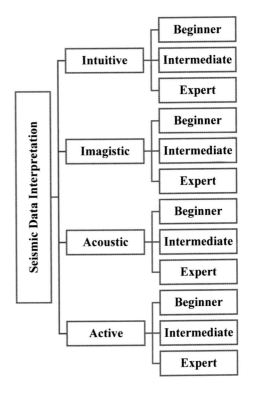

Learning Strategy Level

SeisTutor comprises the total of twelve pedagogy styles that rely on the productive, interactive multimedia techniques, learning styles (learning preferences), and the learning profile (level of difficulty) (see Fig. 4.14).

Course Level

The course level is the hierarchical representation of the Week-Wise learning course material.

Knowledge Concept Level

The subject topics, sub-topics, and their hierarchical representation form a course coverage plan. The course coverage plan also retains the association among the topics and sub-topics (part of, prerequisite).

Henceforth, domain model in SeisTutor gear with the richness of multimedia amalgamated learning materials. SeisTutor retrieves a portion of learning material

Table 4.2 Nomenclature

Symbol	Definition
DKB_{SDI}	Knowledge of seismic data interpretation subject domain
LS_{SDI}	A pool of knowledge capsules or pedagogy style
LT_{SDI}	A collection of learning or subject topics
SST_{SDI}	Set of subject sub-topics
STR_{SDI}	Describes the association between topics
$SSTR_{SDI}$	Defines the association between the sub-topics
ST_{Update}	Update function utilized by the domain or subject experts
SST_{Update}	Subject sub-topic update function utilized by the domain or subject experts

from the knowledge base depending upon the learner's profile and learning style. Furthermore, the learning content is aligned as per the distinct curriculum identified by the SeisTutor (Table 4.2).

Definition A formal representation of a domain or subject knowledge in SeisTutor is defined as follows:

$$DKB_{SDI} = \langle LS_{SDI}, LT_{SDI}, SST_{SDI}, STR_{SDI}, SSTR_{SDI}, ST_{Update}, SST_{Update} \rangle \quad (4.1)$$

$LS_{SDI} = \{PS_1, PS_2, PS_3 \ldots, PS_{12}\}$ It is a pool of knowledge capsules or pedagogy style, depending upon the identified pedagogy style one chosen among them.

$LT_{SDI} = \{LT_1, LT_2, LT_3, \ldots\}$ It is a collection of learning or subject topics covered during the learning session.

$SST_{SDI} = \{ST_1, ST_2, ST_3, \ldots\}$ It is a set of subject sub-topics used to illustrate the topic thoroughly.

$STR_{SDI} = \{STR_1, STR_2, STR_3, \ldots\}$ Subject topic relation describes the association between topics (prerequisite, part of).

$SSTR_{SDI} = \{SSTR_1, SSTR_2, SSTR_3, \ldots\}$ Subject sub-topic relationship defines the association between the sub-topics. Here, association indicates how one sub-topic is associated (prerequisite, part of) with other sub-topics.

ST_{Update}: It is a subject topic update function utilized by the domain or subject experts to perform the required revisions in the course coverage plan.

SST_{Update}: It is a subject sub-topic update function utilized by the domain or subject experts to perform the required revisions in the learning material.

4.3.2.3 Domains of Learning

As per Benjamin Bloom, learning is everywhere, one can use their mental skill, develop an attitude, and acquire physical skills based on their skill used to perform activities [4]. Bloom classifies the domain of learning into three different categories, i.e., Affective, Cognitive, and Psychomotor (see Fig. 4.15).

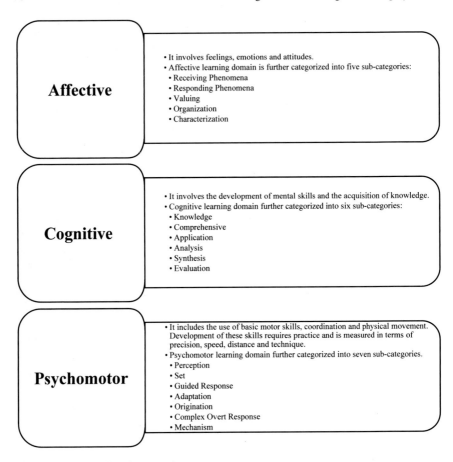

Fig. 4.15 Learning domain categories

In the presented work, four learning styles are taken into the consideration, i.e., Imagistic, Intuitive, Acoustic, and Active [1, 5–8]. The feature of these learning styles is formulated in Table 4.3.

Table 4.3 I2A2 learning styles

Learning style (LS)	Imagistic	Intuitive	Acoustic	Active
Key terminologies	Learning through perceiving	Learning through an understanding of the written word	Learning through hearing	Learning through accomplishment
Interactive multimedia	Flowcharts, diagram, and videos	Written paragraph, written notes, and action charts	Listening	Hand-on exercise

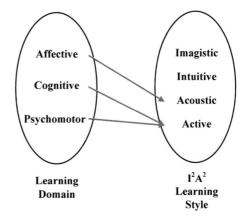

Fig. 4.16 One-to-one mapping between the learning domain and the I2A2 learning style

Figure 4.16 is showing how these learning domains are related to the incorporated learning styles.

4.4 Summary

This chapter presents the distinct features of implicit and explicit knowledge. Establishing the domain of seismic data interpretation, as an experiential domain, it describes the steps involved in the accumulation of tacit knowledge and their characterization and details the transformation of tacit-to-explicit-to-tutor-able. The tutoring material laid into a course delivery pattern comprising 4 weeks and its execution through SeisTutor is described. The content has been organized into topics and sub-topics in the week-wise pattern demonstrated in this chapter. Finally, an adaptive knowledge base is developed that comprises twelve knowledge capsules.

References

1. Ahuja, N.J., Singh, N., Kumar, A.: Development of knowledge capsules for custom-tailored dissemination of knowledge of seismic data interpretation. In: Networking Communication and Data Knowledge Engineering, pp. 189–196. Springer, Singapore (2018)
2. Wu, Y., Liu, W., Wang, J.: Application of emotional recognition in intelligent tutoring system. In: 1st International Workshop on Knowledge Discovery and Data Mining (WKDD 2008), pp. 449–452. (2008)
3. Wolcott, L.: The distance teacher as reflective practitioner. Educ. Technol. **1**, 39–43 (1995)
4. Bloom, B.S., Engelhart, M.D., Furst, E.J., Hill, W.H., Krathwohl, D.R.: Taxonomy of educational objectives: the classification of educational goals. Cognitive Domain, David McKay, New York (1956)
5. Singh, N., Kumar, A., Ahuja, N.J.: Implementation and evaluation of personalized intelligent tutoring system. Int. J. Innov. Technol. Explor. Eng. (IJITEE) **8**, 46–55 (2019)

6. Singh, N., Ahuja, N.J.: Implementation and evaluation of intelligence incorporated tutoring system. Int. J. Innov. Technol. Explor. Eng. (IJITEE) **8**(10), 4548–4558
7. Singh, N., Ahuja, N.J.: Empirical analysis of explicating the tacit knowledge background, challenges and experimental findings. Int. J. Innovative Technol. Exploring Eng. (IJITEE) **8**(10), 4559–4568 (2019)
8. Kumar, A., Singh, N., Ahuja, N.J.: Learning styles based adaptive intelligent tutoring systems: Document analysis of articles published between 2001 and 2016. Int. J. Cogn. Res. Sci. Eng. Edu. **5**(2), 83 (2017)

Chapter 5
Pedagogy Modeling for Building SeisTutor Intelligent Tutoring System

5.1 Introduction

The pedagogy model is the brain of ITS, as it is responsible for making strategic decisions throughout the learning sessions. A strategic decision includes identifying a tutoring strategy, recommending an exclusive course coverage plan, gaging performance parameters, and analyzing the post-tutoring measures (learning gains, learner's emotional state throughout learning, and the degree of understandability). It recommends the course structure, tailoring the representation of learning material depending on the information captured in the learner model. This model comprises three adaptation features: the custom-tailored curriculum sequencing module, tutoring strategy recommendation, and learner performance analyzer module, built into it to facilitate customized tutoring for the learner.

5.2 Workflow of SeisTutor

The implementation is portrayed in various stages (See Fig. 5.1). The pretutoring stage, also characterized by the primary assessment stage, is analyzed below.

Initially, the learner has to create a learner account by registering themselves with the SeisTutor. As soon as a learner account is created, learners are asked to sign in to their account and undergo a pretest. The pretest is the inescapable preassessment test, without which learners are not allowed to proceed with the learning session. This test further opens up the way for the learner to get tutored as per individual learning style and learning level attributes (termed as tutoring strategy).

© The Author(s), under exclusive license to Springer Nature Singapore Pte Ltd. 2022
N. Singh et al., *Cognitive Tutor*, Advanced Technologies and Societal Change,
https://doi.org/10.1007/978-981-19-5197-8_5

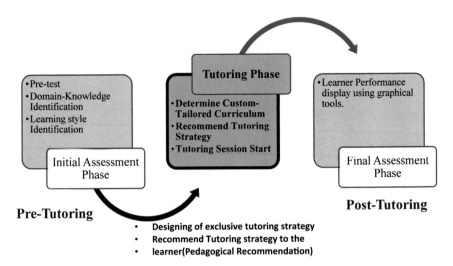

Fig. 5.1 Architecture of SeisTutor

The learner is set up through a pretest with two types of questions: domain knowledge test and learning style test. The questions are broken down into two groups: domain knowledge test comprises 20 problems (questions). The domain knowledge test is designed so that the outcome of this test reveals two characteristics of the learner. One is the learner's competency (indicative of the learner's grasping comfort), and the other is the learner's prior knowledge about the subject domain. The outcome is referred to as "learner level." The learning style test has 18 questions (questions). When people talk about this set, they call it the "learning style question pool." The architecture presented here assists in evaluating the several strategies involved in the instrument under consideration by providing insights. These insights can help generate a clear picture and understanding of the entire process flow, beginning with the initial phase and concluding with the final phase of the accomplishment process. It also includes the tutoring phase, which discusses the tailoring of the curriculum and the methods that must be used to complete each phase effectively.

As per the domain knowledge test scores, the profiles, "beginner," "intermediate," or "expert," are allotted to the learner. After that, the leaner level is alternatively referred to as "learner profile" in this work. The "learning style test" scores are mapped with the learning styles, imagistic, intuitive, acoustic, and active. Thus, once conducted for the learner, the pretest generates output, translated as the learner's learning profile and learning style. This is the output of the pretutoring phase.

The blend of learning style and learner profile is referred to here as pedagogy style. The pedagogy style governs the tutoring mechanism for the learner, and hence, it is termed the tutoring strategy, which is individualistic for the given leaner.

Here is an example of determining what kind of tutoring method to employ based on the student's performance on the pretest.

Assuming the learner's learning style test score is: the imagistic score is 9, for the acoustic score 3, score for the intuitive is 5, whereas for the active is 8, and for the learner's level (as determined by the domain knowledge test scores and hence the allocation of the profiles) is referred as beginner score with 9, the intermediate score with 4 and for the expert score is 7.

SeisTutor maintains a priority list of pedagogical approaches based on test scores and how well they complement one another.

In this case, the learner's level was determined in the pretest as "beginner."

Based on the learning style test findings, pedagogy styles are rated in order of importance. This is the order of imagistic-beginner, active-beginner, intuitive-beginner, and acoustic-beginner. This suggests that the imagistic-beginner pairing is at the top of the list. By offering insights into the instrument's many processes, the planning shown here aids in the evaluation process. They have the capacity to produce a clear picture and knowledge of the entire process flow, from the initial phase to the last phase of accomplishment. There is also a section on tutoring, in which it is made clear that the curriculum must be tailored and the methods utilized to complete each step properly.

In terms of level, they appear to be a beginning, and they favor the imagistic learning style over any other style of learning.

The pedagogical style "beginner + imagistic" should be employed for them.

Further, the learner's prior knowledge, which indicates how much the learner knows before they formally proceed with the learning, is assessed through DKT and is used in designing the custom-tailored curriculum detailed in the next section of this chapter. This diagram of the instrument's architecture aids in the evaluation process by illuminating the many processes. Accordingly, an exclusive tutoring strategy is devised by SeisTutor, with an exclusive course coverage plan comprising the specific topics and sub-topics aligned exclusively for the learner. Their ability to see the big picture and grasp the complete process flow from start to finish makes them ideal candidates for this position. The curriculum and the methods used to complete each step must be suited to the tutoring guidelines provided in the curriculum [1].

5.2.1 Development of Custom-Tailored Curriculum

As per the domain knowledge test scores, the profiles, "beginner," "intermediate," or "expert," are allotted to the learner. After that, the leaner level is alternatively referred to as "learner profile" in this work. The "learning style test" scores are mapped with the learning styles, imagistic, intuitive, acoustic, and active. Thus, once conducted for the learner, the pretest generates output, translated as the learner's learning profile and learning style. This is the output of the pretutoring phase.

The blend of learning style and learner profile is referred to here as pedagogy style. The pedagogy style governs the tutoring mechanism for the learner, and hence, it is termed the tutoring strategy, which is individualistic for the given learner.

Here is an example of determining what kind of tutoring method to employ based on the student's performance on the pretest.

Assuming the learner's learning style test score is: The imagistic score is 9, for the acoustic score 3, score for the intuitive is 5, whereas for the active is 8, and for the learner's level (as determined by the domain knowledge test scores and hence the allocation of the profiles) is referred as beginner score with 9, the intermediate score with 4 and for the expert score is 7.

SeisTutor maintains a priority list of pedagogical approaches based on test scores and how well they complement one another.

In this case, the learner's level was determined in the pretest as "beginner." Based on the learning style test findings, pedagogy styles are rated in order of importance. This is the order of imagistic-beginner, active-beginner, intuitive-beginner, and acoustic-beginner. This suggests that the imagistic-beginner pairing is at the top of the list. By offering insights into the instrument's many processes, the planning shown here aids in the evaluation process. They have the capacity to produce a clear picture and knowledge of the entire process flow, from the initial phase to the last phase of accomplishment. There is also a section on tutoring, in which it is made clear that the curriculum must be tailored and the methods utilized to complete each step properly.

In terms of level, they appear to be a beginning, and they favor the imagistic learning style over any other style of learning.

The pedagogical style "beginner + imagistic" should be employed for them.

Further, the learner's prior knowledge, which indicates how much the learner knows before they formally proceed with the learning, is assessed through DKT and is used in designing the custom-tailored curriculum detailed in the next section of this chapter. This diagram of the instrument's architecture aids in the evaluation process by illuminating the many processes. Accordingly, an exclusive tutoring strategy is devised by SeisTutor, with an exclusive course coverage plan comprising the specific topics and sub-topics aligned exclusively for the learner. Their ability to see the big picture and grasp the complete process flow from start to finish makes them ideal candidates for this position. The curriculum and the methods used to complete each step must be suited to the tutoring guidelines provided in the curriculum (See Fig. 5.2).

After completing the domain knowledge test, the course generator can generate the list of "bugs" wherever the learner provided incorrect responses. The provided module must pass a pretest that examines both knowledge and learning style before it can be presented to students. Their answers are afterward double-checked for accuracy and cross-checked with the topic discussed with respect to domain knowledge [2]. Another feature of this module is that it incorporates a clear visual interface, which is intended to provide exclusive tutoring content so that the learner can rely on

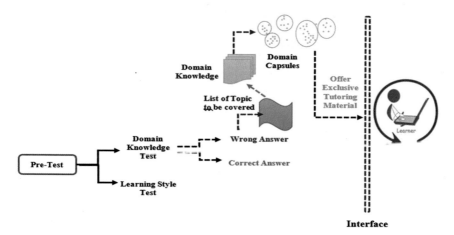

Fig. 5.2 Working of CTCSS module

it for self-clarification and analysis of the work. This list is a collection of topics/sub-topics that learners may not clearly understand [3]. This leads the course generator to recommend the learning path that includes these topics or sub-topics, directing the learner to gain comfort and mastery by offering repetition, re-emphasis, and additional clarity on the subject topics/subtopics the learner currently holds less understanding. This is referred to as custom-tailored curriculum sequencing [4].

5.2.1.1 Mathematical Justification of Custom-Tailored Curriculum Sequencing Module [4]

See Table 5.1.

Let us have ST as the preferred list of the topics to be discussed during the learning session.

$$ST = \{st_1, \ st_2, \ st_3, \ldots \ldots \ldots st_{12}\} \tag{5.1}$$

Let us have DKT_Q as the needed pool of questionnaire enquired in domain knowledge test (DKT) (pretest).

$$DKT_Q = \{dktq_1, \ dktq_2, \ dktq_3, \ldots \ldots \ldots dktq_{12}\} \tag{5.2}$$

Then,

$$DKT_Q \in ST \tag{5.3}$$

Equation 5.3 specifies that each question that is asked during DKT is associated with the list of topics detailed to discuss.

Table 5.1 Nomenclature of custom-tailored curriculum sequrncing module

Symbol	Description
ST	List of topics to be discussed
DKT This is$_Q$	The questionnaire asked in pretest
DKT$_{Correct}$	Correct response
DKT$_{wrong}$	Incorrect response
Curr$_{Topic}$	Topic to be learned
st_1	Introduction
st_2	Seismic waves
st_3	Reflection seismology
st_4	Survey types
st_5	Analog and digital recording equipment's
st_6	Horizon correlation
st_7	Faults in petroleum provinces
st_8	Stress and strain
st_9	Seismic expression of faults
st_{10}	Interpretation of fault data and 3D data
st_{11}	Velocity interpretation
st_{12}	Hydrocarbon indicators

The association between topics and questions can be many-to-one and one-to-one, for the ease of illustration and understanding, one-to-one mapping is considered and demonstrated below (See Fig. 5.3).

Let us have DKT$_{Correct}$ as the list of needed set of questions that are responded correctly, and DKT$_{wrong}$ the list of questions responded incorrectly.

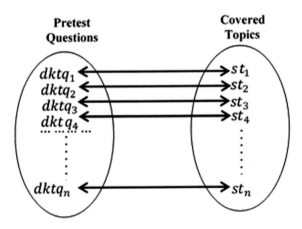

Fig. 5.3 One-to-one type of mapping state

$$DKT_{Correct} = \{dktc_1, \; dktc_2, \; dktc_3, \ldots \ldots \ldots dktc_{12}\} \tag{5.4}$$

$$DKT_{wrong} = \{dktwron_1, \; dktwron_2, \; dktwron_3, \ldots \ldots \ldots dktwron_{12}\} \tag{5.5}$$

It is to be considered that the learner provides accurate responses to (DKT_Q) 20% of the required questions.

$$\text{As } DKT_{correct} \in DKT_Q \text{ and } DKT_{correct} \subseteq DKT_Q \tag{5.6}$$

$$DKT_{correct} = \{dktc_3, \; dktc_6, \; dktc_{10}\} \tag{5.7}$$

Then, it can be considered as 80% of questions DKT_Q fit in to the list of inappropriate responded questions.

$$DKT_{wrong} = \left(DKT_Q - DKT_{correct}\right) \tag{5.8}$$

i.e.,

$$DKT_{wrong} = \{dktw_1, \; dktw_2, \; dktw_4, \; dktw_5, \; dktw_7, \; dktw_8, \; dktw_9, \; dktw_{11}, \; dktw_{12}\} \tag{5.9}$$

$$\text{Similarly, } DKT_{wrong} \in DKT_Q \text{ and } DKT_{wrong} \subseteq DKT_Q \tag{5.10}$$

DKT_{wrong} list is made up of questions that are incorrectly responded by the learner from the pool of questionnaire DKT_Q (See Fig. 5.4).

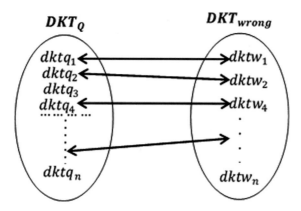

Fig. 5.4 One-to-one mapping between incorrectly responded and domain knowledge questions

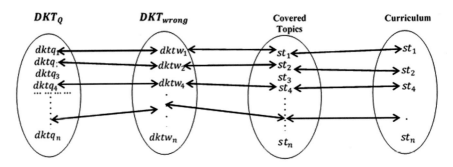

Fig. 5.5 Curriculum designing

Above-mentioned in Eq. 5.3 that $DKT_Q \in ST$.

Therefore by transitivity law of y algebra, i.e., if y Set $X \in Y$, and y Set $Y \in Z$, then $X \in Z$ or

$$X \in Y \in Z$$

$$\text{Thus } DKT_{wrong} \in DKT_Q \in ST \tag{5.11}$$

$$ST = \{set_1, set_2, set_4, set_5, set_7, set_8, set_9, set_{11}, set_{12}\} \tag{5.12}$$

As per Eqs. 5.11 and 5.12, custom-tailored curriculum design is determined by SeisTutor. The mathematical proof is demonstrated in Fig. 5.5. The algorithm for the custom-tailored curriculum sequencing using the bug model is presented below.

5.2.1.2 Algorithm

Input: The outcomes attained of the domain knowledge test

Output: The select curriculum design suitable for the learner
  Begin

1. In a variable, save the answers chosen by the learner for the provided set of questions DKT_Q.

$DKT_{RT} = \{QR_1, QR_2, QR_3, \ldots \ldots \ldots, QR_n\}$. Where RT is a set which encloses, learner's responses, DKT_Q is the set of asked questions, and $QR_1, QR_2, QR_3, \ldots \ldots \ldots, QR_n$ is individual learner responses.

2. A matching procedure is carried out between the obtained results and the real copy of answers kept in the database. AC =

$\{QA_1, QA_2, QA_3, \ldots\ldots\ldots, QA_n\}$. Where AC is a set that enfolds accurate answers and $\{QA_1, QA_2, QA_3, \ldots\ldots\ldots, QA_n\}$ is the corresponding precise solutions.

3.   Distinct the sets of accurate answers $DKT_{Correct}$ and inappropriate answers DKT_{wrong} created.

3.1. do

3.2. {

3.3.   if$(QR_i == QA_i)$

3.4.   {

3.5. $DKT_{correct} = dktq_i;$

3.6.   }

3.7.   else

3.8.   {

3.9. $DKT_{wrong} = dktq_i;$

3.10.   }

3.11. $i++;$

3.12. } while $(i \leq n);$ // where n is the number of questions asked in Preknowledge pretest

4.   Step 2 and Step 3 recurrence for all the answers until while condition is satisfied.

5. DKT_{wrong} sets are looked at and a mapping operation is done between the topics covered and DKT_{wrong} sets.

5.1. for$(k = 1; k \leq n; k++)$

5.2. {

5.3.   for$(l = 1; l \leq n; l++)$

5.4.   {

6.   If$(DKT_{wrong\,k} == QW_k == ST_l)$ // where QW is a set which encloses the labels and $QW = \{QW_1, QW_2, QW_3, \ldots\ldots\ldots, QW_n\}$ are the respective labels associated with the questions.

6.1.   {

6.2.   {

6.3. $ST = st_l;$

6.4.   }

6.5.   else

6.6. {

6.7. continue;

6.8.   }

6.9. }

6.10 }

7.   ST set is made up of the course collection that was made just for the learner by the SeisTutor.

End

5.2.2   Development of Tutoring Strategy Recommendation

As in the case of traditional face-to-face teaching, the content delivery to the learners comprises the human tutor understanding their profile and learning style and accordingly devising the strategy to deliver content. Similarly, the tutoring by SeisTutor is based on key attributes of the learner, the learner profile, learning style, and the prior knowledge assessed through the bug model. The tutoring strategy is termed as custom-tailored curriculum tutoring strategy. This comprises all input parameters obtained from the pretutoring phase. (See Fig. 5.6, Fig. 5.7, and Fig. 5.8). Tables 5.2, 5.3, 5.4, and 5.5 describe the input parameters involved.

Before it can be offered to students, the provided module must pass a pretest that assesses both knowledge and learning style, among other things. Their responses are then double-checked for accuracy and cross-checked with the topic addressed in terms of domain knowledge, learning style, learner profile, and the topic mentioned in terms of domain knowledge. This module also includes a clear comparative method meant to deliver exclusive tutoring content so that the learner can rely on it for custom-tailored strategy and analysis of the job [4, 5].

Twelve combinations of various pedagogy styles have been developed [1, 6, 7]. Each blend is represented as a distinct strategy, and every plan is precharacterized based on the inputs of tutoring parameters for the pretutoring phase. Each combination is mapped with the specific level and style of content to provide to the learners.

The twelve pedagogy styles (PS1, PS2…, PS12), each is a combination of one learner profile and one learning style (See Tables 5.2, 5.3, 5.4, and Fig. 5.6) and is

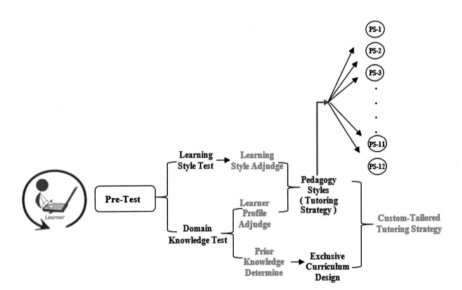

Fig. 5.6 Conceptual flow of tutoring strategy (instructional strategy) computation

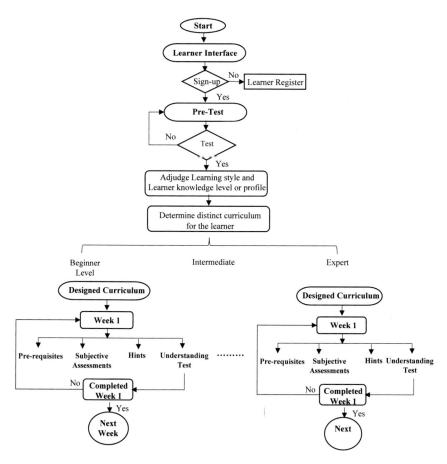

Fig. 5.7 Design of selection of tutoring strategy (TS)

mapped to its content that embodies the said characteristics of respective profile and preferred style of the learner. Based on the learner's prior knowledge, in SeisTutor, as per the algorithm discussed in Sect. 5.2.1.2, alignment of content and designing of an exclusive curriculum (course coverage plan) is done for the learner.

Thus, the developed custom-tailored curriculum tutoring strategy is available to be recommended by the SeisTutor. The design for selecting a tutoring strategy for the different groups of learners is presented in Fig. 5.7.

The tutoring phase, which lasts from Phase 2 to Phase 3, is another phase. During this phase, the student begins the tutoring session, adhering to the tutoring technique specified at the outset. The actions of the learner are also recorded.

We record and analyze psychological and non-psychological factors as part of our work. Learner psychological parameters refer to how a pupil feels during ongoing tutoring sessions. Non-psychological characteristics, on the other hand, are how well

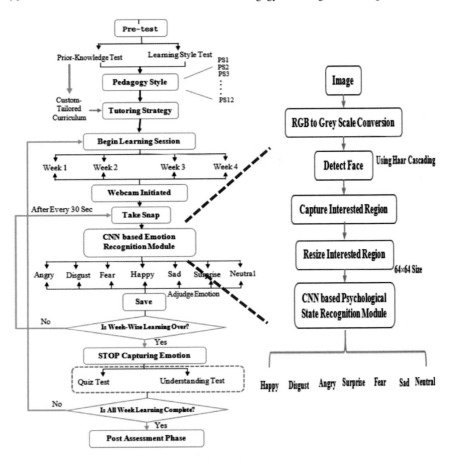

Fig. 5.8 CNN-based emotion states recognition module

Table 5.2 Parameters for tutoring strategy computation

Parameters	Values			
Learner profile (LP)	Beginner	Intermediate		Expert
Learning style (LS)	Imagistic	Intuitive	Acoustic	Active
Prior-knowledge (PK)	List of topics/sub-topics identified based on bugs			

Table 5.3 Learning profile attributes

Parameters	Values		
	Value = 0	Value = 1	Value = 2
Learner profile (LP)	Beginner	Intermediate	Expert
Difficulty level	Easy (simple)	Average (moderate)	Tough (proficient)
Features	More comprehensive illustration	Less comprehensive illustration	Precise (to the point)

Table 5.4 Learning style attributes

Parameters	Values			
	Value = 0	Value = 1	Value = 2	Value = 3
Learning style (LS)	Imagistic	Intuitive	Acoustic	Active
Key terminologies	Learning through perceiving	Learning through an understanding of the written word	Learning through hearing	Learning through accomplishment
Interactive multimedia	Flowcharts, diagram, and videos	Written paragraph, written notes, and action charts	Listening	Hand-on exercise

Table 5.5 Structure of pedagogy styles in "SeisTutor"

S.N.	Pedagogy styles (PS)	Learner profile (LP)	Learning styles (LS)			
			Imagistic	Intuitive	Acoustic	Active
1	PS1	Beginner	✓			
2	PS2	Beginner		✓		
3	PS3	Beginner			✓	
4	PS4	Beginner				✓
5	PS5	Intermediate	✓			
6	PS6	Intermediate		✓		
7	PS7	Intermediate			✓	
8	PS8	Intermediate				✓
9	PS9	Expert	✓			
10	PS10	Expert		✓		
11	PS11	Expert			✓	
12	PS12	Expert				✓

a person performed in the week-by-week evaluation. These include the number of question attempts, correct answers, tips, and the amount of time spent.

At the end of each week, there is usually a checkpoint where the tutoring method can be adjusted (in a user-driven [learner can opt for the flip] or system-driven manner SeisTutor decides and flips based on learner performance measures). It is possible to change the tutoring strategy (flip it) only once during the entire tutoring session. Whether or not to do so during the ongoing tutoring session is decided based on the learner's performance metrics. The choice to switch pedagogy styles is carried out by assigning the next pedagogy style in the priority queue to the current pedagogy style. In other words, when a learner's performance in a current pedagogy style is poor, it is determined that the learner may not be comfortable with the current pedagogy style, and a change in tutoring technique is implemented to improve the learner's performance in that style.

Performance measurements are crucial in measuring the level of comfort a learner feels in a given situation. Their observation of their progress throughout the learning process serves as the foundation for altering the tutoring technique.

Continuous tutoring is grounded on numerical (quantitative) performance measures. The SeisTutor tends to mimic the performance of the human tutor, i.e., by offering the learning material as per the learner's prior or previous knowledge, preferred learning style, and learner grasping levels (obtained in the pretutoring phase), bringing in the adaptive feature. Quantitative values, namely the learner's emotional state, learning gain, and level of understanding, are determined during this type of tutoring.

The third category of the phases involved in the post-tutoring phase. Here, SeisTutor generates the learner progress report, which includes learner week-wise measures for quiz performance, emotions, degree of understanding of learned concepts, session details, time spent on topics, learning gain, and dynamic profile (profile shift (before and after learning)).

5.2.3   CNN-Based Emotion Recognition Model

When the learner begins a new learning session, this module, termed emotion recognition, is activated (shown in Fig. 5.8). As illustrated in Fig. 5.7, as early as the learner in work begins to participate in the learning session, the CNN-based emotion detection module is activated and ready for use, as shown in Fig. 5.8. The CNN-based emotion detection module, based on machine learning techniques such as the convolution neural network, can function properly (CNN). Several studies on facial expression recognition (FER) have found that convolution neural networks are the most effective for things like face position, facial location, and facial scale deviations ([90] and [91]). As a result, a CNN module with nine layers is developed for emotion recognition. Figure 5.8 depicts the operation of the CNN-based emotion recognition module [8–11]. This is how everything works. The emotion recognition module uses a camera to capture an image of the learner's face for analysis. A CNN-based emotion recognition module then processes the image, determining the facial expression of the learner. In future, an examination of the determined emotion will be carried out (Phase 1: evaluation of reaction). Until the student has mastered all of the week's content (themes), the same procedure is repeated repeatedly.

Emotional state, i.e., facial emotion recognition, involves two steps.

1) Face detection
2) Emotion detection

To accomplish this model, it follows two techniques:

Haar cascade classifier: It identifies the frontal face or affected region in an image. There are other techniques to do the same task, but Haar cascade is faster in real-time.

Xception CNN model: For emotion recognition, CNN architecture used in which (48 * 48 pixels) of the bounded face is taken as an input and based on the probabilities, it predicts the emotion.

Dataset

The dataset for training and validation processes is gathered from the Kaggle Website [92]. The dataset contains grayscale face images of (48 * 48 pixels). The primary task is to classify the face image on the basis of seven kinds of emotions (happy = 3, sad = 4, surprise = 5, neutral − , angry = 0, disgust = 1, fear = 2). There are 35,888 samples in the FER2013 dataset.

The following techniques are used for the training of CNN model.

Data Augmentation: This technique is used when the training data is not adequate to learn the image features. It performs operations like normalization, cropping, rotation, zoom, flip, and shear on training dataset.

Kernal Regularizer: During optimization, it puts penalties on layers, these penalties further fused with a loss function. L2 regularization is augmented in CNN.

Batch Normalization: It is used for normalizing the activation of the preceded layer. It is used to increase the speed of the training process.

Global Average Pooling: It is used to reduce feature maps (computing average of all feature map elements) into a scalar value.

Depth-Wise Separable Convolution: It reduces the computational cost (decreasing the number of parameters) in comparison with the standard convolution layer.

5.2.3.1 Case-Wise Response of CNN-Based Emotion Recognition Module

Case 1: Learner created their learner account and signed in to the SeisTutor.

System Behavior: Learner psychological states are not recognized by the CNN-based emotion recognition module immediately after the learner logs on to the learner account.

Case 2: Learner undergoes pretest.

System Behavior: The CNN-based emotion recognition module does not recognize learner psychological states during the preassessment test (pretest).

Case 3: Learner begins the learning session by accessing the lesson under a week (1–4) but does not complete the lesson.

System Behavior: The CNN-based emotion recognition module captures the learner's emotions as the learner accesses lessons for a week (1–4) and finishes by clicking on the "mark as completed" button. The system takes as the learner has completed the lesson. Successively as the learner continues lesson after lesson across weeks, the learner's emotions are captured until the session is active.

Case 4: The learner begins the learning session by accessing the lesson for a week (1–4) but does not finish the lesson, and the session abruptly closes.

System Behavior: If this scenario is encountered, the system does not present the learner's emotions, as no records are maintained.

Case 5: Learner begins with the learning session and continues learning lesson after lesson but does not click the "mark as complete" button.

System Behavior: If this scenario is encountered, the system continues to capture learner emotions until the session is active.

Case 6: Learner begins with the learning session and continues learning lesson after lesson and clicks on the "mark as complete" button of the particular lesson.

System Behavior: If this scenario is encountered, the system captures the learner's emotions for that particular lesson.

Case 7: Learner begins the learning session by clicking on a lesson under a week (1–4) but completes the lesson.

System Behavior: In this scenario, the system captures learner's psychological state for that particular lesson.

Case 8: Learner undergoes performance assessment test (quiz test and degree of understanding module).

System Behavior: Learner's emotions are not captured during the performance assessment test.

5.2.4 Development of Performance Analyzer Model

5.2.4.1 Degree of Understandability

The degree of understandability module aims to identify how effectively a learner understands the taught concepts. The basic flow of this module is shown in Fig. 5.9. The word-based summary analysis is performed to assess in SeisTutor.

The tutoring-based sessions are scheduled and implemented in a pattern that can be covered on a weekly basis. Subsequently completion upon the first week of the learning session, a subjective test has been implemented. This test asks the learner to enter their understanding of the learning content in the form of plain text (See Fig. 5.10). Further, these plain texts are split into sentences using stop words. There is a set of reference matrices, including the main reference matrix and co-occurrence reference matrix. N-gram co-occurrence reference matrix contains the N terms that describe the properties and features of the main reference matrix. Sub-lexicons are in one too many relationships with the main lexicons. Compared with the main reference matrix, learner solutions make a separate matrix of both matched and unmatched text/words. The level similarity measure is quantified (See Fig. 5.10). Now, based on matching lexicons from the main reference matrix, their associated N-gram co-occurrences reference text is retrieved. The next step is to compare the N-gram co-occurrences reference text with the learner solution and make a separate matrix of matched and unmatched text. The N-gram co-occurrence similarity measure is quantified (See Fig. 5.11). From Eqs. 5.30, 5.31, and 5.32, the understanding scores (polarities) or the degree of understanding have been computed. The purpose of

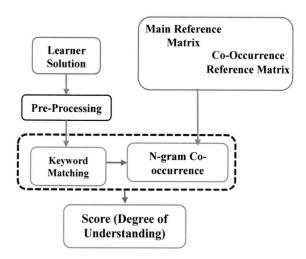

Fig. 5.9 Flow diagram of the performance analyzer module

Fig. 5.10 Step-wise execution of keyword matching technique

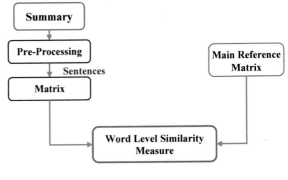

Fig. 5.11 N-gram step-wise execution

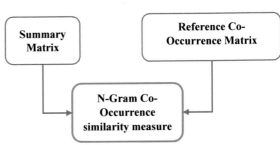

making the unmatched matrix is to give necessary feedback to the learner, which notifies that the learner has not understood or covered these topics and suggests revisiting the specific topics. Based on the understanding score, polarities have been determined.

$$\text{Polarities} = \begin{cases} \text{Above} > 30\,\% \text{ effective impact} \\ \text{Dou} < 30\,\% \text{ negligible impact} \end{cases}$$

Subject : seismic data interpretation

Opinion holder : participants of SeisTutor

5.2.4.2 Mathematical Justification of Performance Analyzer Module

See Table 5.6.

Let us consider MK is the main reference matrix in W_k

$$W = \{W_1, W_2, W_3, W_4\} \tag{5.13}$$

Table 5.6 Nomenclature of performance analyzer module

Symbol	Definition
W	It represents a week, as performance analyzer SeisTutor computes the degree of understanding
MRK	Set of main reference matrix
SK_{mrk_i}	Set of co-occurrence reference matrix subsequence of words that carry in a sentence
X	Learners entered a summary of the learned concepts
ArrMatchList	It contains the words that are in the main reference matrix
ArrResult	It contains the words which are not in both main reference matrix and co-occurrence reference matrix
SubMatchList	It includes the lexicon that is in both main reference matrix and co-occurrence reference matrix
NoTopicList	It contains the words that are not in the main reference matrix
$SMRK_{mrk_1}$	Matched co-occurrence reference matrix
Counter	A variable that counts the rewards when there is a match

$$MRK = \begin{bmatrix} mrk_1 \\ mrk_2 \\ . \\ . \\ . \\ mrk_{10} \\ mrk_{11} \\ mrk_{12} \end{bmatrix} \quad (5.14)$$

Let us consider SK is the co-occurrence reference matrix that illustrates functionality, feature, and property about words in the main reference matrix.

$$SK_{mrk_1} = \begin{bmatrix} mrk_1 \; sk_1 \\ mrk_1 \quad . \\ mrk_1 \; sk_7 \end{bmatrix} \quad (5.15)$$

$$SK_{mrk_2} = \begin{bmatrix} mrk_2 \; sk_1 \\ mrk_2 \quad . \\ mrk_2 \; sk_9 \end{bmatrix} \quad (5.16)$$

$$SK_{mrk_3} = \begin{bmatrix} mrk_3 \; sk_1 \\ mrk_3 \quad . \\ mrk_3 \; sk_5 \end{bmatrix} \quad (5.17)$$

$$\text{Similarly for, } SK_{mrk_{12}} = \begin{bmatrix} mrk_{12}\ sk_1 \\ mrk_{12}\ . \\ mrk_{12}\ sk_8 \end{bmatrix} \tag{5.18}$$

$$SK \in MK \tag{5.19}$$

Equation 5.19 indicates that there is a strong correlation between co-occurrence reference matrix and the main reference matrix, which describes the fruitful information conveyed during learning sessions.

Let us consider the learner entered paragraph, further saved in X for further operation.

$X = \{``\qquad ''\};$

$$X = \begin{bmatrix} x_{11} & \cdots & x_{1n} \\ \vdots & \ddots & \vdots \\ x_{n1} & \cdots & x_{nn} \end{bmatrix} \tag{5.20}$$

Suppose there are four lists named as
ArrMatchmatrix, ArrResultmatrix, SubMatchmatrix, and NoTopicmatrix.
The next operation is to compare matrix X with the main reference matrix.
i.e., compare X with MK.
→ if there is a match, then add MK_i in ArrMatchmatrix.
→ otherwise, add MK_i in NoTopicmatrix.
Suppose out of 12, main reference matrix 3 words are matched.

$$\text{ArrMatchmatrix} = \begin{bmatrix} mrk_1 \\ mrk_4 \\ mrk_6 \end{bmatrix} \tag{5.21}$$

$$\text{NoTopicmatrix} = \begin{bmatrix} mrk_2 \\ mrk_3 \\ mrk_5 \\ mrk_7 \\ mrk_8 \\ mrk_9 \\ mrk_{10} \\ mrk_{11} \\ mrk_{12} \end{bmatrix} \tag{5.22}$$

As aforementioned in Eqs. 5.15, 5.16, 5.17, and 5.18, which indicates that each MK has their own respective SK.

Thus, matched MK is retrieved from ArrMatchmatrix and retrieve their respective co-occurrence reference matrix from SK.

$$SK_{mrk_1} = \begin{bmatrix} mrk_1 \ sk_1 \\ mrk_1 \quad . \\ mrk_1 \ sk_7 \end{bmatrix} \quad (5.23)$$

$$SK_{mrk_4} = \begin{bmatrix} mrk_4 \ sk_1 \\ mrk_4 \quad . \\ mrk_4 \ sk_9 \end{bmatrix} \quad (5.24)$$

$$SK_{mrk_6} = \begin{bmatrix} mrk_6 \ sk_1 \\ mrk_6 \quad . \\ mrk_6 \ sk_5 \end{bmatrix} \quad (5.25)$$

Now compare matrix X with co-occurrence reference matrix, *i.e., compare* ArrMatchmatrix *with* "X".

→ if, there is match has then added SK_{mk_i} in SubMatchmatrix.

→ otherwise, add SK_{mk_i} in ArrResultmatrix.

Suppose matched co-occurrence reference matrix for main reference matrix are as follows

$$SMRK_{mk_1} = \begin{bmatrix} mrk_1 \ sk_1 \\ mrk_1 \ sk_4 \\ mrk_1 \ sk_5 \\ sk_7 \end{bmatrix} \quad (5.26)$$

$$SMRK_{mk_4} = \begin{bmatrix} mrk_4 \ sk_1 \\ mrk_4 \ sk_4 \\ mrk_4 \ sk_5 \\ sk_7 \\ sk_8 \\ sk_9 \end{bmatrix} \quad (5.27)$$

$$SMRK_{mk_6} = \begin{bmatrix} mrk_6 \ sk_1 \\ mrk_6 \ sk_4 \\ mrk_6 \ sk_5 \end{bmatrix} \quad (5.28)$$

$$SubMatchmatrix = \begin{bmatrix} SMRK_{mk_1}, \ SMRK_{mk_4}, \ SMRK_{mk_6} \end{bmatrix} \quad (5.29)$$

Traverse both the ArrMatchmatrix and SubMatchmatrix

$$\text{Word level reference score} = \begin{cases} \text{if match, increment counter by} = 0.5 \\ \text{else,} \qquad \text{counter remain same} \end{cases} \quad (5.30)$$

$$\text{N - gram co - occurrence reference score} = \begin{cases} \text{if match, increment counter by} = 1 \\ \quad \text{else,} \qquad \text{counter remain same} \end{cases}$$

(5.31)

From Eq. 5.30, 5.31, and 5.32, the value of the counter computed below:

$$\text{Counter} = ((3 * 0.5) + (13 * 1))$$

$$\text{Counter} = (1.5 + 13)$$

$$\text{Counter} = 14.5$$

$$\text{DOU} = \left\{ \frac{(\text{Counter})}{\left(\frac{\text{Number of rows in main reference matrix}}{2} \right) + (\text{Number of rows in co - occurrence matrix} * 1)} * 100 \right\}$$

(5.32)

Let us consider the total number of sub-lexicon for 12 main lexica are 40. Then, from Eq. 5.32, the degree of understanding is computed as follows:

$$\text{Degree of understanding} = \left\{ \frac{14.5}{\left(\frac{12}{2} \right) + (40)} \right\} * 100$$

$$\text{Degree of understanding} = \left\{ \frac{14.5}{(46)} \right\} * 100$$

$$\text{Degree of understanding} = 31.52 \%$$

5.2.4.3   Attainment Level

The attainment level of the learner is examined using the "degree of understanding" module. This is the subjective test in which the learner is prompted to enter their understanding of the learning content in the form of plain text. Further, their responses are analyzed to obtain scores and compute the degree of understandability (refer to Eq. 5.30). This score indicates the percentage of learning grasped by the learner.

5.3 Summary

The design of the adaptation modules incorporated in the pedagogy model, the custom-tailored curriculum sequencing model, tutoring strategy recommendation model, and learner performance analyzer module (psychological and non-psychological) has been discussed. Additionally, the composition of the pedagogy style, the design of the course coverage plan depending on prior knowledge, and the methodology of the selection of the tutoring strategy are presented. The tutoring strategy architecture design offers personalized tutoring strategies for educating on a specific topic to the learners.

References

1. Kumar, A., Singh, N., Ahuja, N.J.: Learning styles based adaptive intelligent tutoring systems: document analysis of articles published between 2001 and 2016. Int. J. Cogn. Res. Sci. Eng. Educ. **5**(2), 83 (2017)
2. Singh, N., Kumar, A., Ahuja, N.J.: Implementation and evaluation of personalized intelligent tutoring system. Int. J. Innovative Technol. Exploring Eng. (IJITEE) **8**, 46–55 (2019)
3. Singh, N., Ahuja, N.J.: Implementation and evaluation of intelligence incorporated tutoring system. Int. J. Innovative Technol. Exploring Eng. (IJITEE) **8**(10), 4548–4558
4. Singh, N., Ahuja, N.J., Kumar, A.: A novel architecture for learner-centric curriculum sequencing in adaptive intelligent tutoring system. J. Cases Inf. Technol. (JCIT) **20**(3), 1–20 (2018)
5. Singh, N., Ahuja, N.J.: Bug model based intelligent recommender system with exclusive curriculum sequencing for learner-centric tutoring. Int. J. Web-Based Learn. Teach. Technol. (IJWLTT) **14**(4), 1–25 (2019)
6. Singh, N., Ahuja, N.J.: Empirical analysis of explicating the tacit knowledge background, challenges and experimental findings. Int. J. Innovative Technol. Exploring Eng. (IJITEE) **8**(10), 4559–4568 (2019)
7. Ahuja, N.J., Singh, N., Kumar, A.: Development of knowledge capsules for custom-tailored dissemination of knowledge of seismic data interpretation. In: Networking Communication and Data Knowledge Engineering, pp. 189–196. Springer, Singapore (2018)
8. Singh, N., Gunjan, V.K., Kadiyala, R., Xin, Q., Gadekallu, T.R.: Performance evaluation of SeisTutor using cognitive intelligence-based "Kirkpatrick Model". Comput. Intell. Neurosci. (2022)
9. Singh, N., Gunjan, V.K., Mishra, A.K., Mishra, R.K., Nawaz, N.: SeisTutor: a custom-tailored intelligent tutoring system and sustainable education. Sustainability **14**(7), 4167 (2022)
10. Ahuja, N.J., Singh, N., Kumar, A.: Adaptation to emotion cognition ability of learner for learner-centric tutoring incorporating pedagogy recommendation. Int. J. Control Theory Appl. **9**(44), 15–30 (2016)
11. Singh, N., Gunjan, V.K., Nasralla, M.M.: A parametrized comparative analysis of performance between proposed adaptive and personalized tutoring system "Seis Tutor" with existing online tutoring system. IEEE Access **10**, 39376–39386 (2022)

Chapter 6
Execution of Developed Intelligent Tutoring System

6.1 Implementation of a System

SeisTutor is coded by using the C#.NET framework. Data is stored through the MS Access database running on Windows platform. CNN-based Emotion Recognition Module is implemented using Python, which is integrated into the SeisTutor C# code. This is a standalone offline application compatible to Windows platform.

6.2 Learner Interface Model

The learner interface model is a key component of an ITS because it provides a medium through which a communication is established between the learner and the system. It aids the learner in learning and ITS, for offering learning material during learning sessions. This model not only facilitates the learner to visualize their results but also provides a personalized interaction mode of learning. It plays a significant role in ITS, as it enables the learner to access the system functionalities (See Fig. 6.1). Figure 6.2 presents the main window of the learner interface model.

In addition to this, the learner interface is used to visualize the learning material by means of rich intrinsic multimedia artifacts like pictures/images, audio, and video. By establishing interaction with the other models, this model can offer learning material, assessment materials, hints, feedback, and learning progress statistics to the learner.

6.2.1 Learner Registration

Learner Registration sub-component empowers the learner toward register with the tutoring system. For registration, three credentials are required, a unique username, email id, and password. If the username is unique, then the system saves learner

© The Author(s), under exclusive license to Springer Nature Singapore Pte Ltd. 2022   103
N. Singh et al., *Cognitive Tutor*, Advanced Technologies and Societal Change,
https://doi.org/10.1007/978-981-19-5197-8_6

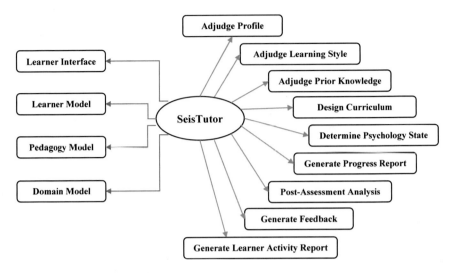

Fig. 6.1 Level 0—DFD of SeisTutor

Fig. 6.2 Main window of the learner interface

credentials and generates a unique learner id. Furthermore, this unique learner id is used to manage the learning sessions and gauge the learner activities, facilitating in making the strategic decisions.

Figures 6.3 and 6.4 illustrate the learner registration process flow and learner registration interface, respectively.

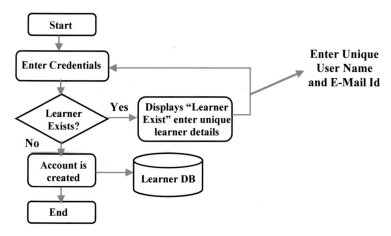

Fig. 6.3 Process of registering with SeisTutor

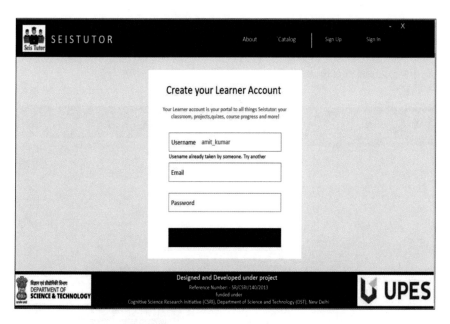

Fig. 6.4 Learner registration interface

Figure 6.3 demonstrates the system which is accepting the details of the learner for the first time; if the learner is existing learner, then the system presents notification to enter unique credentials (Username and Password).

Following security features are built in while creating a new user account. For creating the learner account, the learner has to enter the valid length of username (min 6 character), email id will be an authentic email id, i.e., it should contain one

@ followed by the dot symbol, and password will be in the combination of at least 2 numeric, 1 special character, and 1 capital letter (Figs. 6.5, 6.6, 6.7 and 6.8).

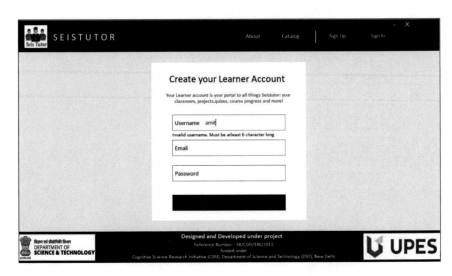

Fig. 6.5 Username validation interface

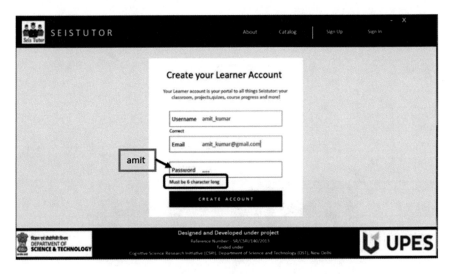

Fig. 6.6 Password character length validation interface

Fig. 6.7   Weak password validation interface

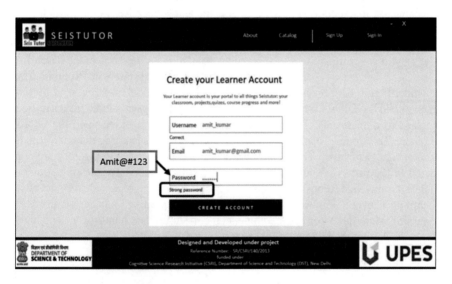

Fig. 6.8   Strong password validation interface

6.3   Domain Model

The domain model comprises the knowledge base of the SeisTutor. It organizes the structure of the course to be delivered (topic/sub-topics and the association between the topics). The domain model represents the "*What-to-teach*" component of the SeisTutor. Figure 6.9 shows the DFD of the domain model.

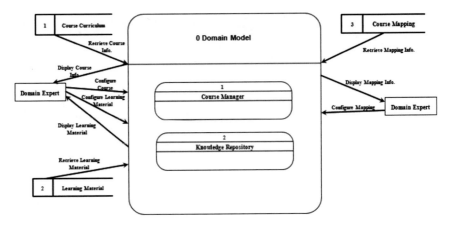

Fig. 6.9 DFD of the domain model

The domain model consists of two sub-components: The first is the Course Manager, which contains the structure of the domain and organizes the course in the manner easy to understand, and the other is the Knowledge Repository, which stores the learning and test material in the database. The learning content is represented by the meta-description attributes. The meta-description forms of learning content help the system to reuse and track the learning content from the knowledge base of the SeisTutor. The experts perform the course mapping task through the expert interface module. Figure 6.10 presents the working of domain model, and Fig. 6.11 presents the domain model interface. As shown in Fig. 6.10, pedagogy model makes retrieval request to domain model, for retrieving learning content based upon the tutoring strategy. The Course Manager of domain model retrieves learning material from the knowledge Repository. Learning Materials in the Knowledge Repository are organized in the form of Knowledge Capsules. As there are twelve tutoring strategies hence, Knowledge Repository comprises twelve Knowledge Capsules. On receiving Learning Content from the Knowledge Repository, Course Manager organizes the Learning Material as per the recommended curriculum and hands over the material to the pedagogy model for further processing [10].

6.4   Learner Model

The learner model is one of the critical components of SeisTutor. It stores the learner characteristics information such as Learner Profile, Learning Style, Prior Knowledge, and Cognitive Skills. SeisTutor aims to provide a personalized learning environment. To accomplish this, SeisTutor incorporates cognitive intelligence. Learner Characteristics (Learner Profile, Prior Knowledge, and Learning Style) play a vital role in

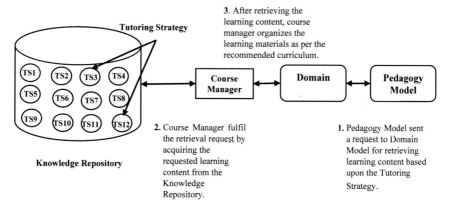

3. After retrieving the learning content, course manager organizes the learning materials as per the recommended curriculum.

2. Course Manager fulfil the retrieval request by acquiring the requested learning content from the Knowledge Repository.

1. Pedagogy Model sent a request to Domain Model for retrieving learning content based upon the Tutoring Strategy.

Fig. 6.10 Step-wise execution of working of domain model

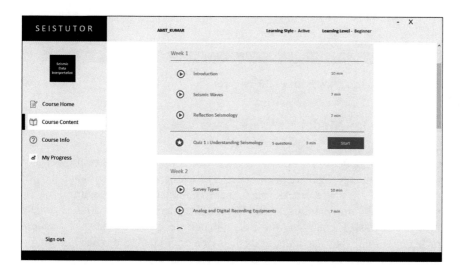

Fig. 6.11 Domain model interface

generating appropriate pedagogy styles, which further improves the learners' performances in many ways. Learning material as per the preferred style of learning makes the learning more comfortable, effective, and adaptable.

The learner characteristic model holds learner characteristics, such as domain Prior knowledge, Learning Style, and Learning Level. The learner characteristics help the system to determine the profile (*Beginner*, *Intermediate*, or *Expert*) of a learner [1, 2]. The screenshot of the learner model interface is presented in Fig. 6.12.

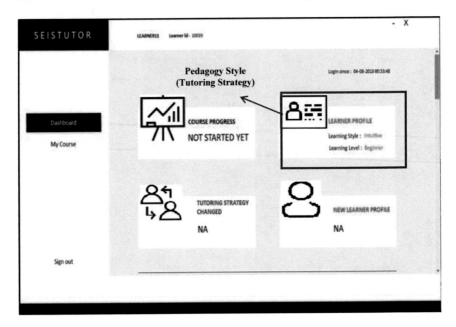

Fig. 6.12 The learner model interface (Home Page)

6.5   Pedagogy Model

The pedagogy model is the brain of an ITS as it is used to make fruitful decisions to mimic the human cognitive intelligence. The learner characteristics and the prior learner knowledge play a crucial role in incorporating adaptation/cognitive features in SeisTutor. As aforementioned, that learner characteristic information is gathered through two tests. The purpose of the domain knowledge test is to discover not only the learner profile, but also identify the learner's prior knowledge (See Fig. 6.13).

Learner Prior Knowledge indicates the acceptable threshold knowledge that the learner is already having on subject topics/sub-topics. Additionally, it helps the Seis-Tutor determine which topics students should focus on most, and these themes are rearranged to create a Custom-Tailored Curriculum for each learner. Curriculum design and teaching style are combined to develop an exclusive tutoring technique. In many aspects, the learner's performance increases as a result of this tailored tutoring technique. SeisTutor not only provides sequenced learning material linked with the learner's comprehension and chosen manner of learning, but also concentrates on the subject topics/sub-topics where the learner is experiencing difficulty. The screenshot of the Custom-Tailored Curriculum offered to the learner dashboard is shown in Fig. 6.15. Figure 6.14 presents the DFD for the pedagogy model and its sub-components.

Fig. 6.13 Domain knowledge test model interface

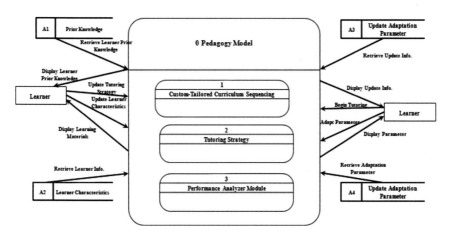

Fig. 6.14 DFD of the pedagogy model

In addition to this, pedagogy model is also responsible for determining the learner's emotions during ongoing learning sessions. Emotion recognition during an ongoing learning session plays a vital role in establishing individuality feature. Here, individuality indicates the individual psychological perceptions. This feature helps the learning systems in determining the learner understanding and overall satisfaction level. Therefore, emotion recognition feature embellishes effective learning sessions and makes the learning session worthwhile [3–6].

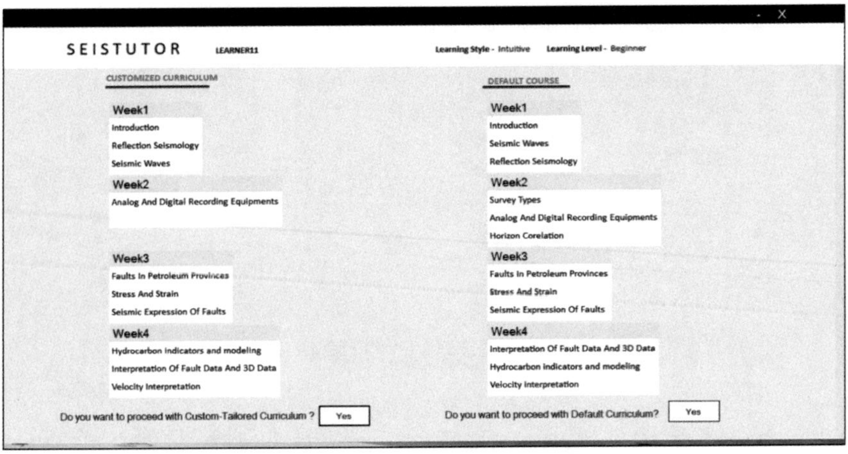

Fig. 6.15 Custom-tailored curriculum offered to the learner

Artificial Intelligence (AI) techniques have been used to develop the Emotion Recognition Module. When a student takes a photo of themselves during a learning session, this module processes the photo and identifies the learner's emotional state (e.g., "Happy" versus "Sad" versus "Angry"). These are then shown alongside the progress of the student. Gathering psychological (emotional) state is repeated until the learner has completed all the content (topics) related with all the weeks of learning. It is now utilized to monitor the emotional condition of students while they are being tutored. The purpose of this module is to identify the learner's emotions toward the learning content offered by the SeisTutor and also identify their perceptions while exploring and learning (overall experience) through SeisTutor. Figure 6.16 shows the working of Emotion Recognition Module, and Fig. 6.17 shows the emotion recognition during ongoing tutoring session.

6.5.1 Performance Analyzer Model

The purpose of the Performance Analyzer module is to identify the degree of understanding of learning content. The degree of understanding is the subjective test, and it is executed in a week-wise pattern. In this test, learner is prompted to enter his/her understanding of the learning content, in the form of plain text. The learner is expected to make use of maximum keywords related to the learning content of the given topics/chapters/sub-topics of the week. The degree of understanding of the lesson is assessed through pattern matching of these keywords with the complete content of that topic/chapter/sub-topic. The degree of understanding formulae is used to quantify the scores and compute the degree of understanding. Figure 6.18 shows the understanding test for a computing degree of understanding.

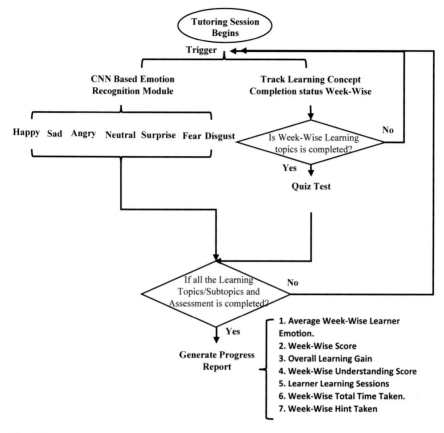

Fig. 6.16 Working of emotion recognition module during tutoring session

6.6 Learner Statistics

The tutoring system maintains two types of data: First is the demographic data that provides necessary information of the learner, such as name, email id, age, highest qualification, and occupation, and second is the personalized learning data that is generated during tutoring session which is further used by the system for decision making. The system records the learner activity during tutoring and also makes use of the personalized data for assessing and evaluating the learner performance such as learner week-wise quiz performance, week-wise maximum occurrence emotions (emotion which was found to be for a maximum period of time during tutoring session), week-wise degree of understanding of learned concepts, login sessions, time spends on topics, Learning Gain, and Dynamic Profile (Profile Shift (before and after learning)) [4, 7–10]. These learner performances are shown with the help of visualization techniques, i.e., pie charts, line diagram, and Bar graph. Figure 6.19 presents the learner progress report generated by the SeisTutor.

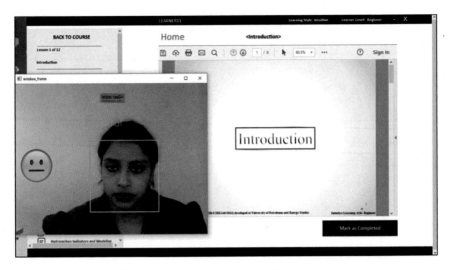

Fig. 6.17 Recognition of emotions during the tutoring session

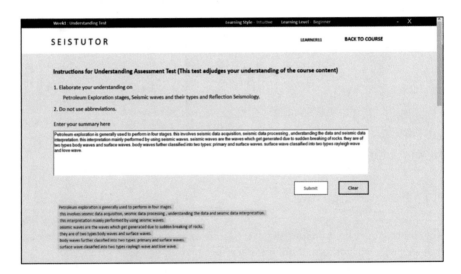

Fig. 6.18 Understanding test for adjudging the learner degree of understanding

6.7 Learner Feedback

The purpose of gathering learner's feedback is to determine the learner's experiences and their perception and provide suggestions for improvement. After completing the learning session, the learner has to give their valuable views. A pool of 44 question questionnaire was created, covering all aspects of SeisTutor such as adaptation features (Custom-Tailored Curriculum, tutoring strategy recommendation, emotion

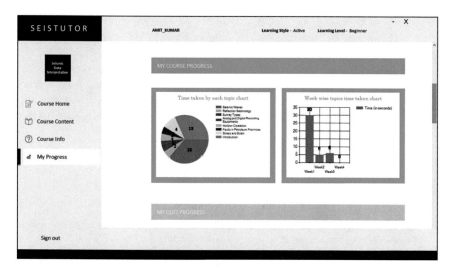

Fig. 6.19 Learning data chart stored with SeisTutor

recognition, and degree of understanding), ongoing support, and personalization feature. Appendix 2 presents the learner feedback report. Learner's feedback plays a vital role in evaluating the effectiveness of the learning process, overall satisfaction level, learner adaptation, and intelligence feature (CTCSS) incorporated in SeisTutor. The Learner feedbacks are ranked under five categories, i.e., strongly satisfied, satisfied, neutral, dissatisfied, and strongly dissatisfied.

This shows that the learner is unable to provide a strong evaluation of their experience with the software. To be strongly unsatisfied and disappointed shows that the student is not content with the features that were used during their learning session. It is a good sign when a learner expresses strong satisfaction and satisfaction with the learning experience. Feedback from students is based on five points, a scale of one to five.

6.8 Summary

This chapter deliberates the execution of learner model, pedagogy model and domain model, Learner Statistics, and Learner Feedbacks components of the SeisTutor. The learner model is developed using fuzzy inference technique. It gauged the learner characteristics and recommended the initial Pedagogy Styles (tutoring strategy). The intelligence features of the pedagogy model are implemented using the "BUG" model (Custom-Tailored Curriculum Sequencing Module), Machine Learning Technique (CNN-based Emotion Recognition Module), and word-based summary analyzer technique (Performance Analyzer Module). The BUG model identifies the learner's previous/prior knowledge by identifying the bugs during the pretest (domain knowledge test) and further recommends the Learner-Centric learning path. A machine learning technique is used to implement the Emotion Recognition Module. A CNN-based Emotion Recognition Module tracks the learner's emotions during ongoing learning sessions. A word-based summary analyzer technique enables the learners to summarize and write their understanding; based on the summary, their understanding score is quantified. The DFD and screenshots of various components of the SeisTutor are shown.

References

1. Ahuja, N.J., Singh, N., Kumar, A.: Development of knowledge capsules for custom-tailored dissemination of knowledge of seismic data interpretation. In: Networking Communication and Data Knowledge Engineering, pp. 189–196. Springer, Singapore (2018)
2. Ahuja, N.J., Singh, N., Kumar, A.: Adaptation to emotion cognition ability of learner for learner-centric tutoring incorporating pedagogy recommendation. Int. J. Control Theory Appl. 9(44), 15–30 (2016)
3. Singh, N., Ahuja, N.J., Kumar, A.: A novel architecture for learner-centric curriculum sequencing in adaptive intelligent tutoring system. J. Cases Inf. Technol. (JCIT) 20(3), 1–20 (2018)
4. Singh, N., Ahuja, N.J.: Bug model based intelligent recommender system with exclusive curriculum sequencing for learner-centric tutoring. Int. J. Web-Based Learn. Teach. Technol. (IJWLTT) 14(4), 1–25 (2019)
5. Singh, N., Gunjan, V.K., Kadiyala, R., Xin, Q., Gadekallu, T.R.: Performance evaluation of SeisTutor using cognitive intelligence-based "Kirkpatrick Model". Comput. Intell. Neurosci. (2022)
6. Singh, N., Gunjan, V.K., Mishra, A.K., Mishra, R.K., Nawaz, N.: SeisTutor: a custom-tailored intelligent tutoring system and sustainable education. Sustainability 14(7), 4167 (2022)
7. Singh, N., Kumar, A., Ahuja, N.J.: Implementation and evaluation of personalized intelligent tutoring system. Int. J. Innovative Technol. Exploring Eng. (IJITEE) 8, 46–55 (2019)
8. Singh, N., Ahuja, N.J.: Implementation and evaluation of intelligence incorporated tutoring system. Int. J. Innovative Technol. Exploring Eng. (IJITEE) 8(10), 4548–4558
9. Kumar, A., Singh, N., Ahuja, N.J.: Learning styles based adaptive intelligent tutoring systems: document analysis of articles published between 2001 and 2016. Int. J. Cogn. Res. Sci. Eng. Educ. 5(2), 83 (2017)
10. Singh, N., Ahuja, N.J.: Empirical analysis of explicating the tacit knowledge background, challenges and experimental findings. Int. J. Innovative Technol. Exploring Eng. (IJITEE) 8(10), 4559–4568 (2019)
11. Singh, N., Gunjan, V.K., Nasralla, M.M.: A parametrized comparative analysis of performance between proposed adaptive and personalized tutoring system "Seis Tutor" with existing online tutoring system. IEEE Access 10, 39376–39386 (2022)

Chapter 7
Performance Metrics: Intelligent Tutoring System

7.1 Overview

This section discusses how statistics were used to evaluate the SeisTutor. Finding out how well the software SeisTutor functions are an important component of the research. The purpose of this research is to determine how successfully SeisTutor assists people in learning about the seismic data interpretation domain in a learner-centered manner. To do this, SeisTutor was tested on a small group of students and teachers from an unnamed university. In total, 60 students volunteered to assist with the evaluation. People were divided into two groups: the control group and the experimental group [1–4].

Control Group Evaluation: Participants in the control group are instructed according to a predetermined curriculum, in which the learning themes and subtopics are presented in the same order from beginning to end. As a result, every single learner follows the exact same course of instruction.

Experimental Group Evaluation: The participants in the experimental group are provided with a custom-tailored curriculum, in which the learning content is realigned depending on the responses of the learners during their domain knowledge test, which was carried out as part of the pretest. In addition, during a learning session, a person's psychological state (emotions) and the degree to which they grasp the material are assessed, both of which are later utilized for the purpose of post-learning analysis.

The aspects of comparison of the both the groups are represented in the (Table 7.1). The purpose to compare the results is to identify the differentiation in the learning experiences.

Both the experiments are executed, and the results have been monitored. Figure 7.3 illustrates the flow of the evaluation process.

The participation of the volunteers in the evaluation process was contingent on their completion of a permission form that provided an overview of the methodology's fundamental components. To take part in the evaluation process, every participant must first receive permission from the organizers [5, 6]. There were a total of 60

© The Author(s), under exclusive license to Springer Nature Singapore Pte Ltd. 2022
N. Singh et al., *Cognitive Tutor*, Advanced Technologies and Societal Change,
https://doi.org/10.1007/978-981-19-5197-8_7

Table 7.1 Feature-wise comparison of SeisTutor performance with two groups

	Control group	Experiment group
Personalized tutoring contents	Offer learning content (similar curriculum) based on tutoring strategy (pedagogy style) (See Fig. 7.1)	Offer personalized learning content (different curriculum) based on adaptive tutoring strategy (See Fig. 7.2)
Psychological state tracking	The emotional state of the learner is not capturing during the ongoing learning session	Determine the emotional state (emotion) of the learner during the ongoing learning session [5]
Degree of understanding computation	Learner's understanding of the concept, not adjudged	Quantify learner's understanding of the concept

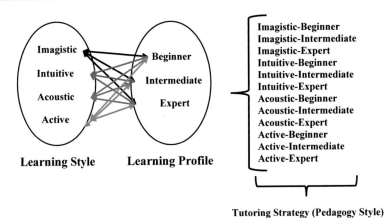

Fig. 7.1 Tutoring strategy (pedagogy style) generation

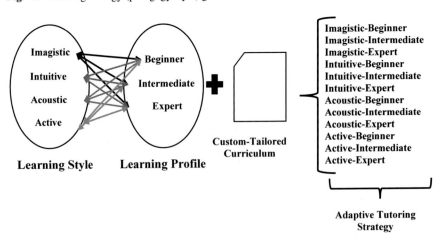

Fig. 7.2 Adaptive tutoring strategy generation steps

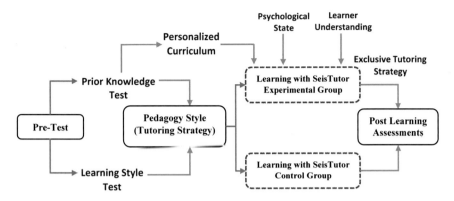

Fig. 7.3 Flow of accomplishing the evaluation process

participants, and 32 of them were assigned to be in the control group. The remaining 28 volunteers were assigned to be in the experimental group. These volunteers will be referred to as either participants or learners in the subsequent sections of the text. The demographic information of the participants is included in Table 7.2.

Throughout the course of the continuing session, SeisTutor will record the learner's pretest scores, quiz scores, feelings, degree of comprehension test scores, and feedback. These are recorded in a distinct manner for the control group and the experimental group. Before continuing with the procedure, these parameters are first normalized. The evaluation consists of two parts: the first part consists of determining which group presents an improvement in learner's learning and aptitudes (i.e., test results), and the second part consists of determining the learner's reactions, comfort level, behavior, and overall results. Both parts of the evaluation are equally important. The first aspects of evaluation are accomplished by using one tailed ANOVA statistical test. The analysis of variance (ANOVA) statistical test is used and considered an appropriate test for judging the significance of the sample means or for judging the significant differences between the two samples (i.e., pretest and the post-test) for both the groups. The analysis of variance (ANOVA) test is used to compare two populations or samples in which there are two instances of observations that match together (e.g., learner test results before and after a specific course, i.e., pretest, post-test). The appropriate test statistics of F-ratio are computed from the sample data, and then, the value based on F-distribution is compared with the result of the calculation (read from the F-table for the different level of significance of the different degree of freedom).

Two estimations of sample variance take into consideration, one based on between samples variance and the other based on within sample variance. Then, both the estimations of sample variance are compared with an F-ratio (See Fig. 7.4).

$$F = \frac{\text{Estimation of population variance depend on between the samples variance}}{\text{Estimation of population variance depend on within the samples variance}}$$

Table 7.2 Demographic characteristics of participating learners

Demographic characteristics			
Characteristic		N = 60	
		Frequency	Percentage (%)
Gender	Male	35	58
	Female	25	42
Age	18–20	7	12
	20–22	11	18
	22–24	11	18
	24–28	3	5
	28–32	7	12
	32–34	13	22
	> 34	8	13
Education	Diploma	0	0
	High/secondary school	18	30
	Graduation	10	17
	Post-graduation	21	35
	Ph.D	11	18
Occupation	Student	18	30
	Teacher	11	18
	Both (teacher and student)	19	32
	Others	12	20

The second aspect of evaluation is accomplished by utilizing Kirkpatrick four phase evaluation model. Kirkpatrick evaluation prototype consists of four-phases shown in Figs. 7.5 and 7.6. Figure 7.7 demonstrates the statistical evaluation technique which is used for demonstrating the evaluation of SeisTutor by employing Kirkpatrick four phase evaluation model.

In addition to this, a comparative analysis is performed between the ITS, i.e., SeisTutor with the existing online learning systems. These learning systems are analyzed on seven parameters, i.e., ("*GUI-based,*" "*learner-centric learning environment,*" "*dynamic profile,*" "the *learning content,*" "the *resolving learner query during session,*" "the *navigation support,*" and "the *learner feedback*").

7.2 Learner Statistics

Table 7.2 shows that men made up 58% of the participants, outnumbering women. The majority of those who took part (35%) have more schooling than a high-school certificate. SeisTutor was created specifically for the subject of "seismic data interpretation." As a result, it is intended for use by students of petroleum engineering

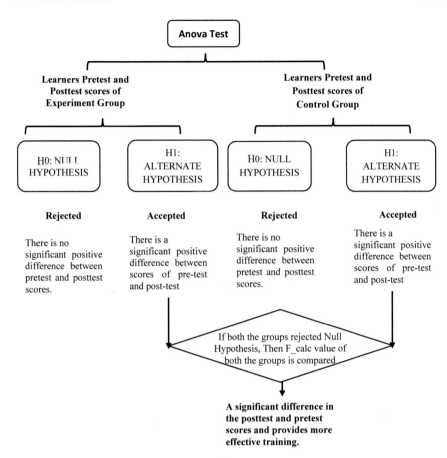

Fig. 7.4 Rules of deduction of ANOVA test

Fig. 7.5 Kirkpatrick's
four-stage evaluation

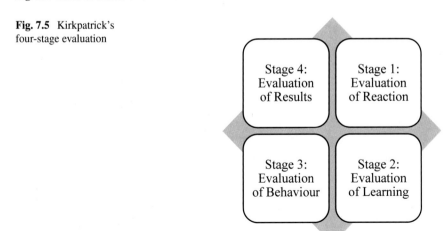

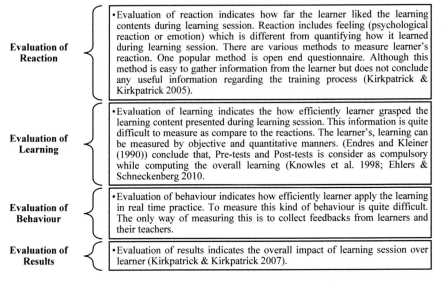

Evaluation of Reaction	• Evaluation of reaction indicates how far the learner liked the learning contents during learning session. Reaction includes feeling (psychological reaction or emotion) which is different from quantifying how it learned during learning session. There are various methods to measure learner's reaction. One popular method is open end questionnaire. Although this method is easy to gather information from the learner but does not conclude any useful information regarding the training process (Kirkpatrick & Kirkpatrick 2005).
Evaluation of Learning	• Evaluation of learning indicates the how efficiently learner grasped the learning content presented during learning session. This information is quite difficult to measure as compare to the reactions. The learner's, learning can be measured by objective and quantitative manners. (Endres and Kleiner (1990)) conclude that, Pre-tests and Post-tests is consider as compulsory while computing the overall learning (Knowles et al. 1998; Ehlers & Schneckenberg 2010.
Evaluation of Behaviour	• Evaluation of behaviour indicates how efficiently learner apply the learning in real time practice. To measure this kind of behaviour is quite difficult. The only way of measuring this is to collect feedbacks from learners and their teachers.
Evaluation of Results	• Evaluation of results indicates the overall impact of learning session over learner (Kirkpatrick & Kirkpatrick 2007).

Fig. 7.6 Brief illustrations of Kirkpatrick's four-stage of evaluation

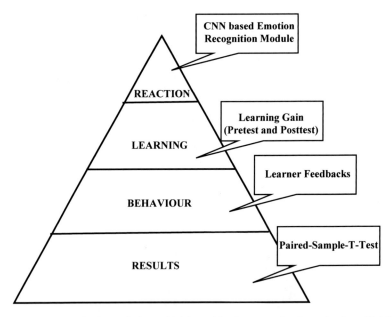

Fig. 7.7 Statistical evaluation technique which is used for demonstrating the evaluation of SeisTutor by employing Kirkpatrick four phase evaluation model

and petroleum exploration. Participants comprised undergraduate petroleum engineering students, teachers from the Department of Petroleum Engineering, and largely practitioners from other professions.

7.2.1 Data Preparation

The data obtained in the experiment was screened through elimination of missing values. Further, normalization of data was performed using min-max normalization. SPSS version 25 was used for the analysis.

7.2.2 Min-Max Normalization

For evaluating the learner's performance, learner's scores (domain knowledge test, post-assessment test, and learning gain) are normalized 0–5 Likert Scale and psychological stats into 0–100. The min-max normalization transforms a value of $X = \{x\_1, x\_2, x\_3, \ldots\ldots, x\_n\}$, fits in the boundary of [A, B]. The expression for min-max normalization is defined below, where

Here, A is considered as the minimum lowest range, and B is considered as the maximum highest range. In this case, the value of [A, B] is [0, 10];

$$Y = \left\{ \frac{x_i - \text{Lowest value in} X}{\text{Highest value in } X - \text{Lowest value in} X} \right\} * (B - A) + A \qquad (7.1)$$

7.3 Learner Performance Metrics

This section assesses the learner's presentation during the education process. The min-max normalization techniques are used to normalize the performance parameters. Three performance parameters are used in this study: pretest score, week-wise quiz scores, and post-test score. The normalization achieves a linear alteration of original marks and turns the score in the range of [0.0, 5.0]. Hence, data range, uniformity is maintained for further processing.

The score of learner performance of week-1, week-2, week-3, and week-4 is calculated and analyzed in both the groups, pretutoring and post-tutoring performance.

7.3.1 Pretutoring and Post-Tutoring Performance

The participant's pretest (pre-tutoring) and post-test performance are quantified and calculated for both the groups. The mean score pretest and post-test for control and experimental groups are 2.41, 3.65, 1.72, and 3.94, respectively. The ANOVA test conducted on these scores [1, 2, 7–9].

H_0: **Null Hypothesis:** There is no noteworthy positive change between pretest and post-test scores, with the post-test score being higher, indicating that there is no improvement.

H_1: **Alternate Hypothesis**: There is a noteworthy positive difference between scores of pretest and post-test, with the post-test score being higher, indicating that there is an improvement.

7.3.2 Predictive Statistical Analysis of Degree of Understanding Module

The computational accuracy, recall, and precision of the degree of understanding module of the performance analyzer module are evaluated in this section. As was just mentioned, after the conclusion of each week, students are required to take quizzes as well as a test to determine their degree of comprehension. During this phase of the evaluation process, the score on the week-by-week quiz is regarded to be the control parameter, while the score on the degree of understanding test is considered to be the predictive parameter. As there are four weeks, this predictive statistical analysis is performed on 28 learners (See Fig. 7.8). Thus, total number of observations are 112 (28 (learners) * 4 (4 weeks understanding test) = 112).

7.3.2.1 Rubrics for Post-Assessment Test

(See Tables 7.3 and 7.4)

7.3.3 Kirkpatrick Four Stage Evaluation (Second Aspects of Evaluation)

7.3.3.1 Kirkpatrick Phase 1: The Evaluation of Reaction

Determined during this stage is the reaction of the student toward the learning content as well as the overall support offered by the educational institution. In order to gage the responses of students, SeisTutor includes both a survey with open-ended

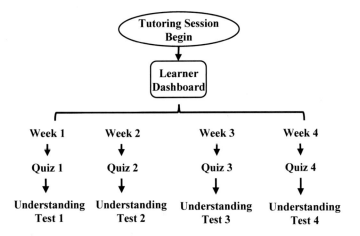

Fig. 7.8 Flow diagram of post-assessment tests

Table 7.3 Performance assessment rubrics for quiz test

Levels	Marks	Description
Level 0	0	Has no understanding of content
Level 1	1	Weak: Struggling to understand the learning content
Level 2	2	Need work: Has little understanding of the material
Level 3	3	Good: Has moderate understanding of the material
Level 4	4	Very good: Has very good or effective understanding of the material
Level 5	5	Excellent: Has perfect or near-perfect understanding of the material

Table 7.4 Performance assessment rubrics for understanding test

Levels	Range	Description
Level 0	(0%)	Has no understanding of content
Level 1	(0–20%)	Weak: Struggling to understand the learning content
Level 2	(20–40%)	Need work: Has little understanding of the material
Level 3	(40–60%)	Good: Has moderate understanding of the material
Level 4	(60–80%)	Very good: Has very good or effective understanding of the material
Level 5	(80–100%)	Excellent: Has perfect or near-perfect understanding of the material

questions and a CNN-based reaction recognition unit (learner feedback). Therefore, in this evaluation, learner's emotions during ongoing learning sessions take into the consideration (participants of experimental group) (See Fig. 7.9).

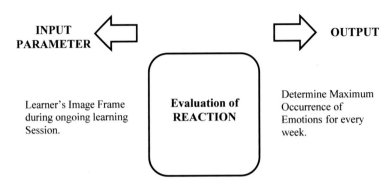

Fig. 7.9 Input–output parameter for the evaluation of REACTION

7.3.3.2   Kirkpatrick Phase 2: Evaluation of Learning

In this phase, learners overall learning is quantified. Therefore, learners quiz and degree of understanding test scores take into the consideration. In this phase, learning gain for both the groups is quantified (quiz scores) and compared (See Fig. 7.10).

Furthermore, a correlation analysis, i.e., bivariate Pearson correlation is performed on these data. This correlation analysis determines the relationship between two parameters. The hypothesis test for P value is.

H0: There is no significant relationship between quiz score and degree of understanding score.

Ha: There is a statistically significant relationship between quiz score and degree of understanding score.

Another reason of performing this correlation analysis is that this degree of understanding test is conducted only for the experimental group of learners. Therefore, the statistical comparative analysis based on the degree of understanding scores is not

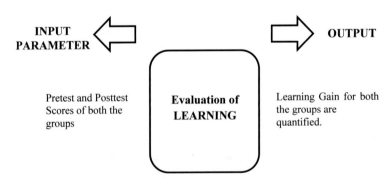

Fig. 7.10 Input–output parameter for the evaluation of LEARNING

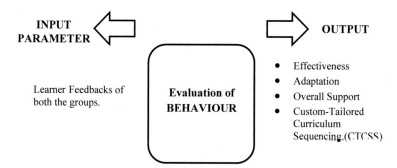

Fig. 7.11 Input–output parameter for the evaluation of BEHAVIOR

possible to perform. If this test rejects, the H0: null hypothesis, then there a linear relationship between learning gain and degree of understanding score.

7.3.3.3 Kirkpatrick Phase 3: Evaluation of Behavior

In this phase, learner's behavior toward effectiveness, adaptation (incorporated artificial intelligence features), overall support, learner comfort level, and custom-tailored curriculum sequencing is quantified (See Fig. 7.11). Therefore, learner feedbacks are taken into the consideration. Learner feedbacks are taken on a five point Likert scale 0–5 (strongly satisfied, satisfied, neutral, dissatisfied, and strongly dissatisfied).

7.3.3.4 Kirkpatrick Phase 4: Evaluation of Results

This step quantifies overall results in terms of effective learning. As a result, the pretest and post-test scores of individuals in both studies (experimental and control groups) are taken into account. Paired sample T-test is used to measure the effectiveness of learning. The paired sample T-test is a robust T-test that assesses if the mean difference between the pretest and post-test scores is zero or not. A high-mean change among pretest and post-test scores is required for effective learning. (See Fig. 7.12). Here, two cases take into consideration.

Case 1: The paired-sampled-T-test implemented on experimental group.

Hypothesis-Case-1.0:

Assume that the pre and post-test mean scores of the participants in the experimental group are the same (insignificant performance improvement).

Hypothesis-Case-1.1:

Let the participants in the experimental group become mixed up with differing mean scores from the pre and post-tests (effective performance improvement).

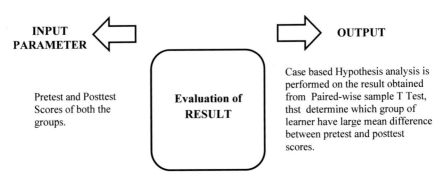

Fig. 7.12 Input–output parameter for the evaluation of RESULT

Case 2: The paired-sampled-T-test performed on control group.

Hypothesis-Case-2.0:

Participants in the control group should have mean scores on the pretest and post-test that are similar (negligible performance improvement).

Hypothesis-Case-2.1:

Participants in the control group should have pretest and post-test mean scores that differ from those of the study participants (effective performance improvement).

7.4   SeisTutor: A Comparative Analysis with Teachable, My-Moodle and Course-Builder Learning Management System

7.4.1   My-Moodle

My-Moodle is a free and open-source tutoring system that may be utilized by academic researchers who want to set up a learning environment and test their intelligent teaching system. Classes hosted on my-Moodle make available a wide array of resources and tools, such as glossaries, assignments, quizzes, databases, and additional tools. The goal of my-primary Moodle is to provide activity-based modeling, in which tasks are broken up into sequences that lead the learner along a learning path. Figure 7.13 is a picture of the dashboard for my-Moodle.

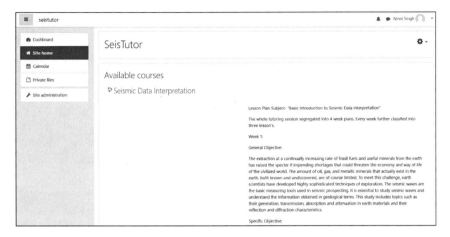

Fig. 7.13 My-Moodle dashboard

7.4.2 Course-Builder

Course builder lets researchers make their own learning environments, such as subject domains and learning quizzes, without having to learn how to code. Since Google App Engine is used to build the course-builder, there is no limit to how many people can sign up to take the courses. It helps keep the connection between the teacher and the students. Their goal is to make education available to as many people as possible, which is why they work with Openedx. The course builder dashboard is shown in Fig. 7.14.

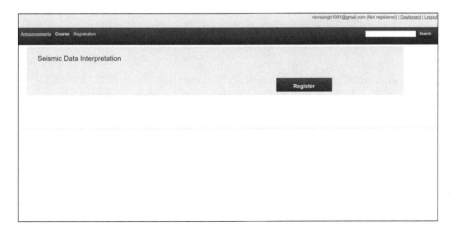

Fig. 7.14 Course-builder dashboard

Fig. 7.15 Teachable dashboard

7.4.3 Teachable

Teachable is an open-source tutoring tool that streamlines the educational process and makes learning easy. It gives subject matter experts a place to submit their own learning content, regardless of the technology that was used to develop it. Teachable LMS is user-friendly, contributes to the growth of your brand, and is an excellent choice for those who own their own businesses. However, they did not place an emphasis on personalizing and adapting the learning experience to the level of comprehension and preferred medium of each individual student. Figure 7.15 depicts teachable dashboard.

7.4.4 SeisTutor

The SeisTutor acts in a manner that is analogous to that of the human instructor. The purpose of SeisTutor is to provide a learning environment that is custom-tailored for tacit knowledge in the subject domain of "seismic data interpretation." As a result, it makes use of components from the fields of computer science and artificial intelligence, such as the custom-tailored curriculum sequencing module, the adaptive tutoring strategy recommendation module, and the performance analyzer module (CNN-based Emotion Recognition Module and degree of understandability module). Learner-centric learning material is what SeisTutor provides to its customers. This means that the learning material is aligned according to the learner's preferred learning style, learning profile or level, and prior knowledge level [7, 8]. During ongoing tutoring, performance parameters, such as degree of engagement, emotions, quiz score, and learning gain, are determined. Figures 7.16 and 7.17 represent the learner dashboard.

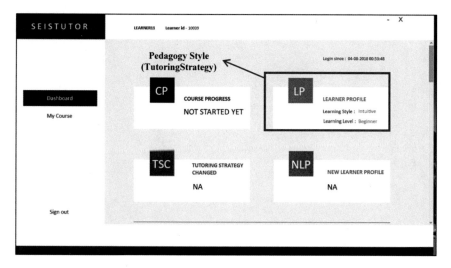

Fig. 7.16 Learner dashboard using SeisTutor

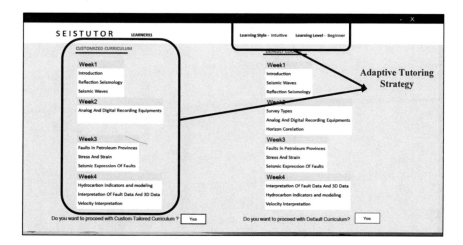

Fig. 7.17 Custom-tailored curriculum offered to the learner

The comparison here is based on how each tutoring system operates, as indicated in Table 7.5. Seventy people have signed up to learn about "seismic data interpretation." Teachable, course-builder, my-Moodle, and SeisTutor were all reviewed by the same group of 70 students, and their useful comments are presented [3, 4, 10, 11].

Feedback can be classified into three categories: severely unhappy, neutral, and strongly satisfied. The learner cannot rate their experience with the system as positive or negative. Strongly displeased indicates that the student did not enjoy using the feature throughout the learning session. The learner is strongly satisfied if they are

Table 7.5 Summary of exiting tutoring system

Parameters		My-Moodle (LMS)	Course builder	Teachable	SeisTutor	Conclusion drawn
GUI-based	Interactive GUI	Yes	Yes	Yes	Yes	All learning systems had done the significant work on the looks and felt of the tutoring system, which further helps to learn and made the system easy to use. Feedback report indicates that 71% of learner are highly satisfied with the SeisTutor GUI in comparison with my-Moodle (47%), course builder (50%), and teachable (57%)

(continued)

Table 7.5 (continued)

	Parameters	My-Moodle (LMS)	Course builder	Teachable	SeisTutor	Conclusion drawn
Learner-centric learning environment	Learner learning style	No	No	No	Yes	Learner-centric learning environment is an essential aspect of the tutoring system because it makes the learning effective. But, as per the feedback received, it has been noticed that my-Moodle, course builder, and teachable have given less emphasis on determining learner preferences learning style and custom-tailored curriculum as compared to SeisTutor. However, some significant amount of work has been done by the course builder for identifying curriculum but still lacking behind to adapt the learner's prior knowledge and provide the personalized learning path. Feedback report indicates that 82% of learner are highly satisfied with the SeisTutor GUI in comparison with my-Moodle (38%), course builder (4%), and teachable (4%)
	Learner learning level	No	No	No	Yes	
	Adaptive tutoring strategy	No	No	No	Yes	
	Custom-tailored curriculum	No	Separate course tracks features are there, but they are not customized. They are pre-decided by the administrator based on the advance, and basic course opted by the learner	No	Yes	

(continued)

Table 7.5 (continued)

	Parameters	My-Moodle (LMS)	Course builder	Teachable	SeisTutor	Conclusion drawn
Dynamic profiling	Learner pretest	No	No	No	Yes	SeisTutor carefully analyzes the learner psychological state and performance parameter before beginning learning session, after learning session and during the learning session. Thus, based on his/her interaction with the system SeisTutor dynamically analyze (learning gain) and update the learner profile. While other learning systems analyze the learner after the completion of the learning session. Feedback report indicates that 83% of learner are highly satisfied with the SeisTutor GUI in comparison with my-Moodle (46%), course builder (27%), and teachable (23%)
	Learner post-test	Yes	Yes	Yes	Yes	
	Learner psychological state during ongoing session	No	No	No	Yes	

(continued)

Table 7.5 (continued)

	Parameters	My-Moodle (LMS)	Course builder	Teachable	SeisTutor	Conclusion drawn
Learning content	Passive learning contents	Yes	Yes	Yes	No	SeisTutor offers learning material in total twelve pedagogy style, while my-Moodle, course-builder, and teachable offer learning material only in one style. Feedback report indicates that 78% of learner are highly satisfied with the SeisTutor GUI in comparison with my-Moodle (14%), course builder (16%), and teachable (3%)
Resolving learner query during session	Handle learner problem during the session	No	Yes	Yes	No	SeisTutor and my-Moodle are unable to handle learner runtime issues, while teachable and course-builder offer learners issue at runtime. Feedback report indicates that 68% of learner are highly satisfied with the SeisTutor GUI in comparison with my-Moodle (34%), course builder (28%), and teachable (38%)

(continued)

Table 7.5 (continued)

	Parameters	My-Moodle (LMS)	Course builder	Teachable	SeisTutor	Conclusion drawn
Navigation support	Other parameters (navigation, modality, language, learning goal)	Yes	Yes	Yes	Yes	All learning system had done the significant work on providing the excellent navigational support in the tutoring system, which further helps to navigate from one module to other easily. Feedback report indicates that 68% of learner are highly satisfied with the SeisTutor GUI in comparison with my-Moodle (51%), course builder (50%), and teachable (60%)
Learner feedback	Learner feedback	Yes	Yes	Yes	Yes	All learning system captures the learner feedbacks to analyze the effectiveness and comfort level of the learning system. Feedback report indicates that 65% of learner are highly satisfied with the SeisTutor GUI in comparison with my-Moodle (53%), course builder (46%), and teachable (64%)

pleased with the feature they used during the learning session. Learners can provide feedback on a Likert scale ranging from 1 to 5.

7.5 Summary

This chapter describes the statistical methods used for the evaluation of the SeisTutor. In the following chapter, the results and findings through the evaluation of SeisTutor are discussed.

References

1. Singh, N., Kumar, A., Ahuja, N.J.: Implementation and evaluation of personalized intelligent tutoring system. Int. J. Innovative Technol. Exploring Eng. (IJITEE) **8**, 46–55 (2019)
2. Singh, N., Ahuja, N.J.: Implementation and evaluation of intelligence incorporated tutoring system. Int. J. Innovative Technol. Exploring Eng. (IJITEE) **8**(10), 4548–4558
3. Singh, N., Gunjan, V.K., Kadiyala, R., Xin, Q., Gadekallu, T.R.: Performance evaluation of SeisTutor using cognitive intelligence-based "Kirkpatrick Model". Comput. Intell. Neurosci. (2022)
4. Singh, N., Gunjan, V.K., Mishra, A.K., Mishra, R.K., Nawaz, N.: SeisTutor: a custom-tailored intelligent tutoring system and sustainable education. Sustainability **14**(7), 4167 (2022)
5. Ahuja, N.J., Singh, N., Kumar, A.: Adaptation to emotion cognition ability of learner for learner-centric tutoring incorporating pedagogy recommendation. Int. J. Control Theory Appl. **9**(44), 15–30 (2016)
6. Ahuja, N.J., Singh, N., Kumar, A.: Development of knowledge capsules for custom-tailored dissemination of knowledge of seismic data interpretation. In: Networking Communication and Data Knowledge Engineering, pp. 189–196. Springer, Singapore (2018)
7. Singh, N., Ahuja, N.J., Kumar, A.: A novel architecture for learner-centric curriculum sequencing in adaptive intelligent tutoring system. J. Cases Inf. Technol. (JCIT) **20**(3), 1–20 (2018)
8. Singh, N., Ahuja, N.J.: Bug model based intelligent recommender system with exclusive curriculum sequencing for learner-centric tutoring. Int. J. Web-Based Learn. Teach. Technol. (IJWLTT) **14**(4), 1–25 (2019)
9. Singh, N., Ahuja, N.J.: Empirical analysis of explicating the tacit knowledge background, challenges and experimental findings. Int. J. Innovative Technol. Exploring Eng. (IJITEE) **8**(10), 4559–4568 (2019)
10. Singh, N., Gunjan, V.K., Nasralla, M.M.: A parametrized comparative analysis of performance between proposed adaptive and personalized tutoring system "Seis Tutor" with existing online tutoring system. IEEE Access **10**, 39376–39386 (2022)
11. Kumar, A., Singh, N., Ahuja, N.J.: Learning styles based adaptive intelligent tutoring systems: document analysis of articles published between 2001 and 2016. Int. J. Cogn. Res. Sci. Eng. Educ. **5**(2), 83 (2017)

Chapter 8
Analysis of Performance Metrics

8.1 Critical Analysis of Performance Metrics

This analysis determines the overall learner performance. Learner performance (scores) in quizzes (week-1 week-2, week-3, and week-4) is analyzed, and the average of the same is computed for both the groups, i.e., experimental and control.

Tables 8.1 and 8.2 show the average week-wise learner's performances of both the groups (control and experiment), respectively.

Control Group
The mean score performance of week-1 is [3.22], week-2 is [3.63], week-3 is [3.74], and week-4 is [4.00] as shown in Fig. 8.1. The performance graphs of learner's for week-1, week-2, week-3, and week-4 of the control group are shown in Figs. 8.2, 8.3, 8.4 and 8.5, respectively.

Experimental Group
The mean score performance of the week-1 is [3.43], week-2 is [3.93], week-3 is [4.07], and week-4 is [4.32] as shown in Fig. 8.6. The performance graphs of learner's for week-1, week-2, week-3, and week-4 of the experimental group are shown in Figs. 8.7, 8.8, 8.9 and 8.10 respectively.

When the average performance of each learner across both groups is examined on a week-by-week basis, the results demonstrate that the experimental group outperforms the control group from the first week forward through the fourth week, while the control group's performance remains relatively constant. The mean score improvements of week-1 is [0.21], week-2 is [0.3], week-3 is [0.32], and week-4 is [0.32]. From their mean score week-wise improvement, one can say with confidence that learning through adaptive tutoring strategy helps the learner to enhance the learner performance (see Fig. 8.11).

Table 8.1 Overall performance of control group

S. No.	Before tutoring		During tutoring				After tutoring	
	Pretest score	Learner level (LL)	Week-wise learner performance				Post-tutoring score	Learner level (LL)
			W1	W2	W3	W4		
L1	2.31	INT	3	2	3	4	3	INT
L2	2.31	INT	1	3	4	4	3	INT
L3	4.62	EXP	4	4	5	4	4.25	EXP
L4	2.31	INT	2	3	2	5	3	INT
L5	4.62	EXP	3	4	5	4	4	EXP
L6	0.38	BEG	4	4	3	4	3.75	EXP
L7	2.31	INT	3	3	4	5	3.75	EXP
L8	2.31	INT	2	4	4	5	3.75	EXP
L9	4.62	EXP	5	5	5	4	4.75	EXP
L10	2.69	INT	3	4	3	4	3.5	INT
L11	0.38	BEG	1	3	4	5	3.25	INT
L12	2.31	INT	4	4	4	5	4.25	EXP
L13	2.31	INT	2	3	5	4	3.5	INT
L14	2.31	INT	3	2	3	4	3	INT
L15	1.54	BEG	3	3	4	4	3.5	INT
L16	2.69	INT	4	5	4	5	4.5	EXP
L17	2.31	INT	3	4	3	4	3.5	INT
L18	2.31	INT	4	5	5	4	4.5	EXP
L19	0	BEG	3	4	4	3	3.5	INT
L20	2.69	INT	2	3	4	3	3	INT
L21	2.31	INT	2	3	3	4	3	INT
L22	0	BEG	3	3	3	4	3.25	INT
L23	2.15	INT	5	5	5	4	4.75	EXP
L24	2.31	INT	4	4	3	3	3.5	INT
L25	2.31	INT	3	3	3	4	3.25	INT
L26	0.38	BEG	4	5	4	3	4	EXP
L27	4.62	EXP	5	3	4	3	3.75	EXP
L28	2.31	INT	4	4	4	4	4	EXP
L29	4.62	EXP	4	4	4	4	4	EXP
L30	4.75	EXP	4	5	4	5	4.5	EXP
L31	3.5	INT	2	2	2	3	2.38	INT
L32	0.38	BEG	4	3	3	3	3.25	INT
Avg	2.41		3.22	3.63	3.75	4	3.65	

Table 8.2 Overall performance of experiment group

S. No.	Before tutoring		During tutoring				After tutoring	
	Pretest score	Learner level (LL)	Week-wise learner performance				Post-tutoring score	Learner level (LL)
			W1	W2	W3	W4		
L1	2.69	INT	2	3	3	5	3.25	INT
L2	1.54	BEG	3	3	4	5	4	EXP
L3	1.92	BEG	4	4	5	3	4	EXP
L4	2.69	INT	5	4	4	4	4.25	EXP
L5	1.54	BEG	3	3	4	4	3.5	INT
L6	1.92	BEG	3	4	5	4	4	EXP
L7	0.38	BEG	4	5	3	5	4.25	EXP
L8	1.54	BEG	5	5	3	5	4.5	EXP
L9	1.15	BEG	5	4	4	3	4	EXP
L10	2.31	INT	3	3	4	5	3.75	EXP
L11	0.38	BEG	2	3	3	4	3	INT
L12	2.31	INT	2	3	4	5	3.5	INT
L13	1.15	BEG	5	5	3	4	4.25	EXP
L14	0.77	BEG	4	3	4	5	4	EXP
L15	1.15	BEG	3	5	3	4	3.75	EXP
L16	1.92	BEG	2	4	4	4	3.5	INT
L17	2.31	INT	3	4	5	5	4	EXP
L18	1.54	BEG	3	3	4	4	3.5	INT
L19	0.13	BEG	3	4	5	3	3.75	EXP
L20	0.77	BEG	2	3	5	5	3.75	EXP
L21	1.92	BEG	5	4	3	5	4.25	EXP
L22	3.43	INT	3	5	4	4	4	EXP
L23	0.77	BEG	5	4	5	3	4.25	EXP
L24	0.38	BEG	5	4	5	4	4.5	EXP
L25	4.23	EXP	3	5	4	5	4.25	EXP
L26	3.08	INT	4	5	5	5	4.75	EXP
L27	1.54	BEG	3	5	4	4	4	EXP
L28	2.69	INT	2	3	5	5	3.75	EXP
Avg	1.72		3.43	3.93	4.07	4.32	3.94	

The performance of the student on the pretest is used by SeisTutor to classify the learner into one of three levels, namely beginner, intermediate, or expert. Based on the results of the learners in the control group's pretest, it was discovered that initially, 7 of the learners belonged to the beginner category, 19 of the learners belonged to the intermediate category, and 6 of the learners belonged to the expert category. Learners'

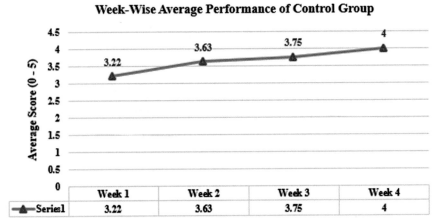

Fig. 8.1 Average learner performance of control group

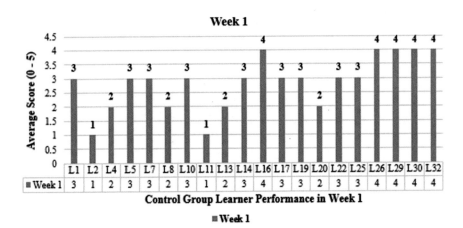

Fig. 8.2 Control group learner performance in week-1

initial profiles are either raised in quality or lowered in quality according on how well they performed in general during their tutoring sessions. From the results, it has been observed that, 7 learners initially belonged to the beginner profile, out of which 4 get promoted to intermediate and remaining 2 get promoted to the expert profile (see Table 8.3).

In a similar manner, initially, there were 19 learners who fit the profile of being intermediate. It has been observed that out of 19 learners, 12 learners continue to remain at the same level (beginner), while 7 of them get promoted to the expert profile. This is the case. Six learners initially belonged to the expert profile. Of these six learners, five learners continue to remain at the same level (expert), and only one learner is demoted to the intermediate profile.

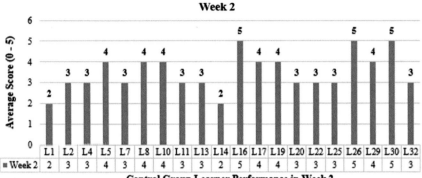

Fig. 8.3   Control group learner performance in week-2

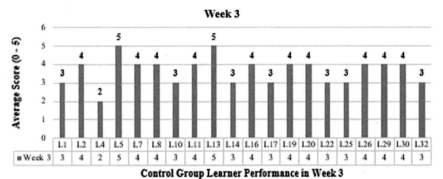

Fig. 8.4   Control group learner performance in week-3

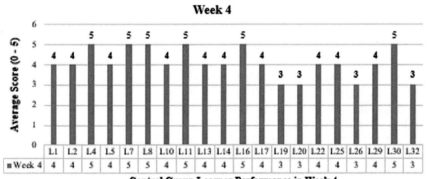

Fig. 8.5   Control group learner performance in week-4

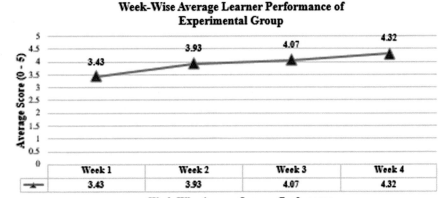

Fig. 8.6   Average learner performance of experimental group

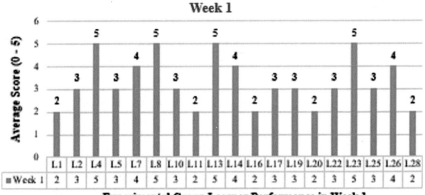

Fig. 8.7   Experimental group learner performance in week-1

When the results of the second phase's pretest were taken into consideration, it was discovered that initially, there were 19 students who belonged to the beginner category, 8 students who belonged to the intermediate category, and one student who belonged to the expert group.

Initially, 19 learners belonged to *beginner* profile out of them 4 get promoted to *intermediate* and remaining 15 get promoted to *expert* profile (see Table 8.4). In a similar manner, there were initially 8 learners who belonged to the intermediate profile; out of these 8 learners, only 2 learners continue to be at the same level, while the remaining 6 learners get promoted to the expert profile. In the beginning, there was just one learner who belonged to the expert profile, and that learner is still at the same level.

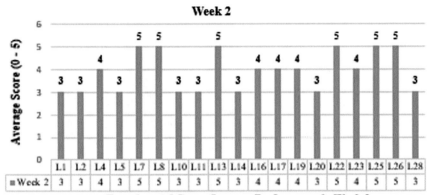

Fig. 8.8 Experimental group learner performance in week-2

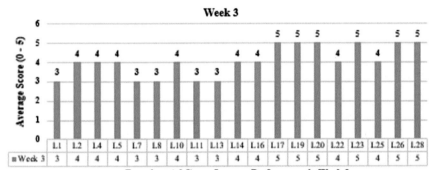

Fig. 8.9 Experimental group learner performance in week-3

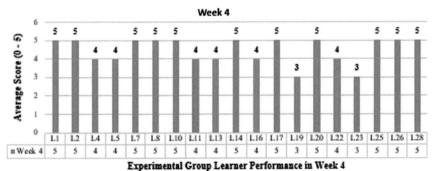

Fig. 8.10 Experimental group learner performance in Week-4

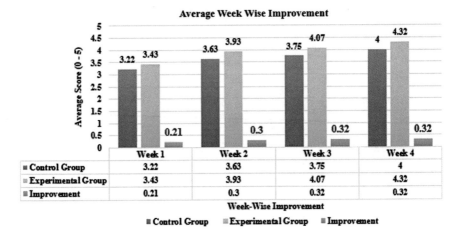

Fig. 8.11   Average week-wise Improvement

Table 8.3   Learner level before and after tutoring (control group)

S. No.	Before tutoring		After tutoring		
	Learner level	Learners	Beginner	Intermediate	Expert
1	Beginner	7	0	4	3
2	Intermediate	19	0	12	7
3	Expert	6	0	1	5

Table 8.4   Learner level before and after tutoring (experimental group)

S. No.	Before tutoring		After tutoring		
	Learner level	Learners	Beginner	Intermediate	Expert
1	Beginner	19	0	4	15
2	Intermediate	8	0	2	6
3	Expert	1	0	0	1

Therefore, this can be concluded from the aforementioned analysis, that the learners, which belongs to experimental group, improved their performance in terms of scores (shown in Figs. 8.12 and 8.13). Here, learner profile upgradation from their initial learning profile specifies the effectiveness of the learning program, i.e., dynamic profiling of the learner.

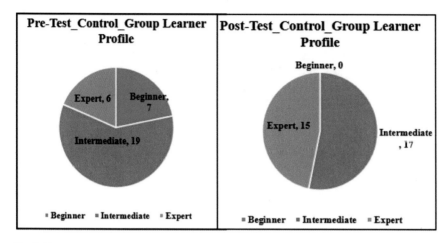

Fig. 8.12 Learner level before and after tutoring (control group)

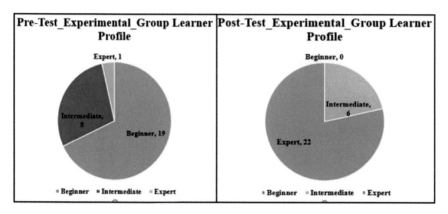

Fig. 8.13 Learner level before and after tutoring (experimental group)

8.2 Statistical Analysis of Learner Engagement

This section discusses the results and finding of the predictive analysis performed on the scores obtained by the learner in the degree of understanding test.

Table 8.5 summarizes "degree of understanding computations" using 2×2 confusion matrix that portrays four possible circumstances.

Let first quantify the precision–recall and accuracy.

Precision: Proportion of positive cases predicted positively.

$$\text{Precision} = \frac{\text{TP}}{\text{TP} + \text{FP}} \tag{8.1}$$

Table 8.5 2 × 2 confusion matrix

True positive Condition: effective impact correctly predicted	False positive Condition: negligible impact incorrectly predicted
False negative Condition: effective impact incorrectly predicted	True negative Condition: negligible impact correctly predicted

Recall: Proportion of positive cases predicted accurately (consider all the cases).

$$\text{Recall} = \frac{\text{TP}}{\text{TP} + \text{FN}} \tag{8.2}$$

Accuracy: Proportion in which both positive and negative cases predicted accurately.

$$\text{Accuracy} = \frac{\text{TP} + \text{TN}}{\text{TP} + \text{FN} + \text{FP} + \text{TN}} \tag{8.3}$$

Let the threshold value is 2, then Table 8.6 depicts the 2×2 confusion matrix.

From Eqs. 8.2, 8.3, and 8.4, the computed precision, recall, and accuracy are as follows:

$$\text{Precision} = \frac{94}{94} =\gg 1 \tag{8.4}$$

$$\text{Recall} = \frac{94}{94 + 15} =\gg 0.8623 \tag{8.5}$$

$$\text{Accuracy} = \frac{94}{94 + 15 + 03} =\gg 0.84 \tag{8.6}$$

Similarly, when the threshold value is 3, then Table 8.7 depicts the 2×2 confusion matrix.

From Eqs. 8.1, 8.2, and 8.3, the computed precision, recall, and accuracy are as follows:

Table 8.6 2 × 2 confusion matrix when the threshold value is 2

True positive	False positive 0
94	0
False negative	True negative
150	3

Table 8.7 2×2 confusion matrix when the threshold value is 3

True positive 75	False positive 0
False negative 19	True negative 18

$$\text{Precision} = \frac{75}{75} =\gg 1 \tag{8.7}$$

$$\text{Recall} = \frac{75}{75 + 19} =\gg 0.7978 \tag{8.8}$$

$$\text{Accuracy} = \frac{75}{75 + 19 + 18} =\gg 0.67 \tag{8.9}$$

From Eqs. 8.6, 8.7, 8.8 and 8.9, it has been observed from the obtained results are that when the threshold value increases, the recall and accuracy value are decreasing, but the precision value remains the same. Precision and recall are inversely proportional to each other. The results from Eqs. 8.4, 8.5 and 8.6, the prediction accuracy of the degree of understanding module is 84%. It may get varied based on the threshold value.

8.3 Pretutoring and Post-tutoring Performance

This section identifies and discusses the results of the groups that prompts enhancements in learner's learning/aptitudes (i.e., test results). The ANOVA is performed on the pretest and post-test score of learners of both the groups. The inference on the obtained results is as follows.

The F-ratio of the ANOVA test for the control group is $F\_calc = 21.68911$ at $\alpha = 0.05$, where α is a significant level, while the tabulated value is $F\_\alpha = 4.00$ (from the F-table). Here, $F\_calc > F\_\alpha$, with the degree of freedom being $v1 = 1$ and $v2 = 62$ (see Table 8.10). The learning gain has been shown in Table 8.8. Hence, the null hypothesis H_0 is rejected, and the alternative hypothesis, $H_1 : \mu 1 < \mu 2$ is accepted. It indicates that there is a significant difference between scores of pretest and post-test tests. Hence, it is deduced that the difference in the post-test and pretest is significant, and the training is effective for the control group.

Table 8.8 Data of pretest and post-test in terms of learning gain

System	Total participants	Pretutoring test score	Post-tutoring test Score	Mean learning gain (%)
Experimental group	28	1.72	3.94	44.4
Control group	32	2.41	3.65	24.8

Table 8.9 Data of pretest and post-test in terms of learning gain experimental group

Source of variation	SS	DF	VS	F-ratio	5% F-limit (from the F table)
Between sample	68.86446	$(2 - 1) = 1$	68.86446	119.7141	$F(1, 54) = 4.03$
Within sample	31.06302	$(56 - 2) = 54$	0.575241		
Total	68.86446	$(56 - 1) = 55$			

Table 8.10 Data of pretest and post-test in terms of learning gain control group

Source of variation	SS	DF	VS	F-ratio	5% F-limit (from the F table)
Between sample	24.88763	$(2 - 1) = 1$	24.88762656	21.68911	$F(1, 62) = 4.00$
Within sample	71.1432	$(64 - 2) = 62$	1.147470917		
Total	96.03082	$(64 - 1) = 63$			

Similarly, the F-ratio of the ANOVA test for the experimental group is $F\_calc = 119.7141$ at $\alpha = 0.05$, where α is a significant level, while the tabulated value is $F\_\alpha = 4.03$ (from the F-table). Here, $F\_calc > F\_\alpha$, with the degree of freedom being $v1 = 1$ and $v2 = 54$ (see Table 8.9). Hence, the null hypothesis Ho is rejected, and the alternative hypothesis H_1: $\mu 1 < \mu 2$ is accepted. It indicates that there is a significant difference between pretest and post-test tests. Hence, it is deduced that the difference in post-test and pretest is significant, and the training is effective for the experimental group. The learning gain has been shown in Table 8.8.

The participant's performance for both the phases reject the null hypothesis, which means training provided in both the groups is effective. However, the aims to identify that which phase or group of training has a higher impact on enhancing the overall learning gain. Participants in the experimental group will be given personalized instruction as part of the program. Sequenced learning material (based on their prior or previous knowledge) and learner facial expressions are captured (ongoing learning session), and the degree of the understanding score is determined, whereas the control group participants do not have access to these features enabled by the learning environment. Therefore, to conclude, F_calc of both the groups is compared. F_calc of experimental group ($119.7141 - 21.68911 = 98.02499$) is higher than F_calc of the control group. Therefore, compared to the control group, the experimental group delivers more effective training and reports a significant difference in the post-test and pretest scores [1–6].

8.4 Learner Learning Analysis Using Evaluation Model "Kirkpatrick"

This section identifies, discusses, and compares the obtained results based on the 4 phases of the Kirkpatrick model, namely learner's reactions, comfort level, behavior, and overall results [7]. A four phase evaluation is performed on learner's emotion, pretest score, post-test scores, quizzes, and feedback that belonged to both the groups. The phase-wise inference on the obtained results is as follows.

8.4.1 Kirkpatrick Phase 1: Evaluation of Reaction

The emotion recognition module, which is based on CNN, is used to determine how the learner feels about the content being presented. The original scores are subjected to a linear transformation before being normalized using the min-max method, which then places the values into the range [0–10]. The normalization of the scores is done so that they all remain consistent with one another.

The emotions of the learner are determined only for the participants, who have been involved in the experimental group evaluation. Thus, the average mean score percentage of maximum emotion occurrence is shown in Table 8.11. The result of 28 participants is shown in Table 8.11, in which 44% of emotion are happy, population of 40% of emotion are considered as neutral, 36% of emotion are considered as angry, 32% of emotion are considered as surprise, 30% of emotion are fear, and 24% of emotion is sad. Thus, from the result, the maximum emotion observed is happy (44%), which specifies that the 44% of learners are happy with the provided learning content and teaching process (pedagogy) [7–10].

Table 8.11 Descriptive statistics of psychological parameter of the learner for experimental group

Emotions	Mean	Std. deviation	Mean (%)
Happy	4.4174	29.6357	44.1
Sad	2.4272	24.9175	24.2
Surprise	3.2275	28.2939	32.2
Fear	3.0612	26.8571	30.6
Angry	3.6728	26.1069	36.7
Neutral	4.0389	26.6193	40.3

8.4.2 Kirkpatrick Phase 2: Evaluation of Learning

This phase quantified the learner's overall learning, i.e., learning gain. Equation 8.10 is used for computing learning gain. The inferences on the results are as follows.

$$\text{Learning\_Gain} = (\text{PostTest\_Score}_L - \text{PreTest\_Score}_L) \qquad (8.10)$$

The average learning increase for the experimental group is 44.34%, while it is only 24.8% for the control group. So, it can be said that the SeisTutor successful in increasing learner curiosity, which indirectly ultimately increases learning gain, if the learning materials are available based on the student's preferences, and a curriculum is made just for them based on what they already know. (see Table 8.12).

Table 8.12 describes the progressive learning gain of 44.34% among learners that participated in the experimental group. Furthermore, this data (learning gain) is used for correlation analysis, i.e., bivariate Pearson correlation. This test is performed between the learning gain and the degree of understanding score of experimental group. As described in Sect. 4, the performance analyzer module is implemented for determining the degree of understanding. But this module is not offered to the control group. Therefore, this correlation analysis aims to determine the correlation between learning gain and the degree of understanding score. If there is a correlation, then from the law of symmetry, i.e., $if, A \in B$ and $B \in C$ then, $A \in C$, one can say with confidence that if this test is offered to the control group, then in that case also, the participants of the experimental group having a higher degree of understanding score (against the control group) (see Table 8.13).

The learning gain by itself has a correlation of 1, which can be explained by the fact that one of the variables or parameters is perfectly correlated with itself. The Pearson correlation between learning gain and degree of understanding is 0.484, and the significance level for this correlation is two-tailed, meaning that the P-value is less than 0.05. As a result, it has been determined that the degree of understanding

Table 8.12 Learner's learning gain

Study cases	No of participants (n)	Learning gain		
		Mean	Standard deviation	Mean (%)
Experiment group	28	2.2170	1.02795	44.34
Control group	32	1.2793	1.37034	24.8

Table 8.13 Average mean score of learning gain and degree of understanding

Parameters	No. of participants (n)	Learning gain		
		Mean	Standard deviation	Mean (%)
Learning gain	28	2.2170	1.02795	44.34
Degree of understanding	28	2.5467	1.31201	50.9

and the amount of knowledge gained have a linear connection that is statistically significant. ($P < 0.05$) (see Fig. 8.14 and Table 8.14).

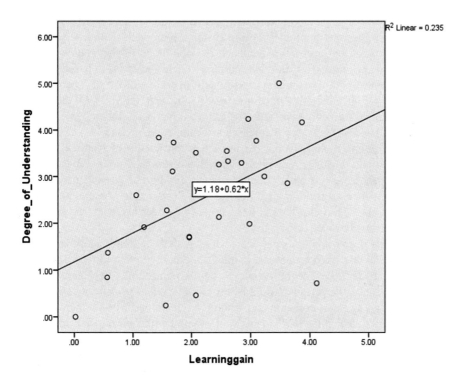

Fig. 8.14   Linear relationship with learning gain and degree of understanding

Table 8.14   Correlation matrix between learning gain and degree of understanding

Parameters		Learning gain	Degree of understandings
Learning gain	Pearson correlation	1	0.484[**]
	Sig. (2-tailed)		0.009
	N	28	28
Degree of understanding	Pearson correlation	0.484[**]	1
	Sig. (2-tailed)	0.009	
	N	28	28

8.4.3 Kirkpatrick Phase 3: Evaluation of Behavior:

Based on learner feedback, this phase quantifies/assesses the learner's behavior toward, effectiveness, adaptation (incorporated artificial intelligence features), overall support, learner comfort level, and custom-tailored curriculum sequencing.

SeisTutor asks the learner for feedback as soon as all of the week's learning concepts have been finished. In this section, a conclusion is made based on what the learners have said. The students are thought to be one of the best parts of this evaluation. Overall, about 93% of people were happy with SeisTutor. Of those, 45% were very happy, and 48% were happy as well (see Fig. 8.15, Table 8.15). With the SeisTutor, it has also been seen that learners became more productive.

22 questions were asked about the impact of the intelligent features provided by SeisTutor, and the same has been collected and summarized in Table 8.16. As some intelligent features are not provided to the control group participants (custom-tailored curriculum sequencing module, Emotion Recognition Module, and degree of understanding). Thus, feedbacks of 28 learners have been taken into consideration from experimental group participants.

Most of the people who took part were happy with the system's adaptive tutoring strategy, which was liked by 86% of them. Of these, 46% were satisfied, and 40% were very satisfied. 85% of the people who took part thought that learning from their own mistakes helped them do a better job. Forty-five percent of respondents expressed high levels of contentment, while the remaining forty-five percent were content overall. 85% of the participants were pleased with the exclusive curriculum that was claimed by the system. Of these participants, 35% were satisfied with the

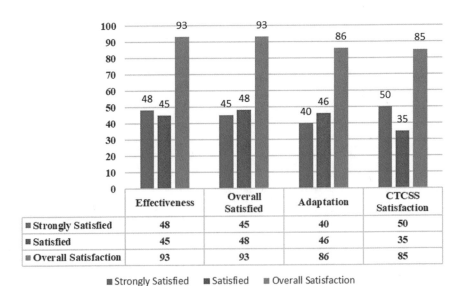

	Effectiveness	Overall Satisfied	Adaptation	CTCSS Satisfaction
■ Strongly Satisfied	48	45	40	50
■ Satisfied	45	48	46	35
■ Overall Satisfaction	93	93	86	85

■ Strongly Satisfied   ■ Satisfied   ■ Overall Satisfaction

Fig. 8.15 Evaluation of learner behavior based on learner feedbacks

Table 8.15 Learner feedback on effectiveness of SeisTutor

	Questions	Degree				
		Strongly satisfy	Satisfy	Neutral	Dissatisfy	Strongly dissatisfy
System Effectiveness	What is your overall level of satisfaction with SeisTutor?	27	23	6	3	1
	The learning through this tutoring system (SeisTutor) was easy	27	26	5	1	1
	Did you feel that you were achieving learning outcomes?	30	21	6	3	0
	I would recommend a course through SeisTutor with no instructor help	29	24	3	5	0
	Would you recommend SeisTutor to individual who needs to take another course?	25	27	5	3	0
	Did SeisTutor support you to make your study productive?	28	27	3	0	2
	How well does this system deliver on your learning intentions?	31	21	5	2	1

curriculum, and 50% were very satisfied with it. 92% of participants agreed that the understanding exam given at the end of each week should correspond to the material covered in that week's lectures. Of those participants, 39% strongly agreed, and the remaining 53% agreed as well. At last, 82% participants agreed that CNN-based emotion recognition module accurately determined their emotions during learning, in which 39% were strongly agreed, and 43% were satisfied as well.

Table 8.16 Learner feedback on adaptation of SeisTutor

	Questions	Degree				
		Strongly Satisfy	Satisfy	Neutral	Dissatisfy	Strongly dissatisfy
Adaptivity/personalization	Did SeisTutor satisfy you with dynamic creation of your learning profile?	27	21	7	1	4
	Were you convenient and satisfied with the tutoring strategy presented to you by SeisTutor?	24	19	9	6	2
	The information provided by SeisTutor is at a level that you understand	29	17	12	0	2
	The tutoring session was at the right level of difficulty for you	26	23	9	2	0
	As a learner, did you feel that your learning style was appropriately judged?	29	25	3	2	1
	Once, tutoring begins and you were tutored, were your learning preferences sufficiently satisfied?	24	24	9	2	1

(continued)

Table 8.16   (continued)

Questions	Degree				
	Strongly Satisfy	Satisfy	Neutral	Dissatisfy	Strongly dissatisfy
Did the experience of learning by your own learning preference, make you perform better?	24	21	6	7	2
Based on your prior subject knowledge, has SeisTutor accurately determined exclusive curriculum for you?	14	8	2	3	1
How satisfied are you with the exclusively determined curriculum?	13	7	4	3	1
As a learner did you feel learning material enabled you to improve your ability to formulate and analyse the problem?	10	14	1	3	0
Are you satisfied with the sequencing of learning content?	14	09	3	2	0

(continued)

Table 8.16 (continued)

	Questions	Degree				
		Strongly Satisfy	Satisfy	Neutral	Dissatisfy	Strongly dissatisfy
	Does sequencing of learning material relate with your previous knowledge? (Give Rating)	12	11	2	3	0
	Has this learning session been successful in improving your knowledge in the subject domain? (Give Rating)	12	11	2	3	0
	Did this learning material fulfill your expectations?	11	13	2	1	1
	The Understanding Test at the end of each week corresponds to the lessons taught?	11	13	2	0	2
	SeisTutor compels and supports me to complete the quizzes, understanding test and lessons ?	13	12	2	0	1

(continued)

Table 8.16 (continued)

	Questions	Degree				
		Strongly Satisfy	Satisfy	Neutral	Dissatisfy	Strongly dissatisfy
	The post tutoring evaluation system (week wise understanding) as it exists is:	14	10	1	2	1
	How do you rate the sequence of the lessons in the course?	18	8	0	0	2
	Has SeisTutor accurately determined your psychological (emotional) state during tutoring session? (Give Rating)	11	7	5	5	0
	Do you feel recognition of emotion during ongoing tutoring is indicative of empathy of the system?	13	12	2	0	1
	The course content are relevant and well organized ?	14	10	1	2	1

The 60 people who filled out the learner's feedback questionnaire helped figure out how well SeisTutor helped the learning process as a whole (see Table 8.17). The results of the analysis showed that 87% of the participants are happy with SeisTutor as a whole, with 47% saying they are satisfied and 40% saying they are very satisfied. Also, 78% of the participants are happy with how the system helps them find the information they need, with 43% saying they are satisfied and 35% saying they are very satisfied.

Table 8.17 Learner feedback on SeisTutor ongoing learning support

Questions	Degree				
	Strongly Satisfy	Satisfy	Neutral	Dissatisfy	Strongly dissatisfy
How are you satisfied with the system support?	24	17	11	6	2
The system navigation support enabled finding the needed information easily	21	17	9	11	2
Was the pre-learning procedure available in SeisTutor helpful to you?	25	17	6	9	3
Were you able to understand the language used to explain the lessons in SeisTutor?	33	21	6	0	0
The tutoring was flexible to meet my learning requirements	30	21	7	2	0

Table 8.18 demonstrates how useful various aspects of the learning content are to the process of learning. These aspects include topic descriptions, revisions, quizzes, and question hinting. 85% of students indicated that they were pleased with how SeisTutor explained the material for the course, with 47% indicating that they were satisfied and 38% indicating that they were extremely satisfied. In addition, 78% of the student body showed interest and concurred that the available tutoring resources were enough; 35% of the student body was extremely satisfied, and 43% of the student body was satisfied. Evidently, both the tests and the counsel were genuine and concentrated on the topics that SeisTutor planned to instruct on.

The overall evaluation of SeisTutor, which was based on student input, found that 86% of students thought that tutoring was delivered in accordance with their learning profile or level, learning style, and background knowledge. The students or participants were quite interested in the artificial intelligence capabilities such as the automatic recommendation of adaptive tutoring strategy, dynamically measuring learner performance, emotion recognition, and evaluation of the degree to which the learner understands.

The feedback from the learners was retrieved and looked at in a free form way. Some students put their ideas on how to make SeisTutor work better. Most of the suggestions were general and had to do with improving the system. Eight of the suggestions were negative when it came to improving the quality of the learning materials, the video lessons, and the hints that the system gave. In the end, 87%

Table 8.18 Learner feedbacks on learning material, quizzes, and overall SeisTutor support

Questions	Strongly Satisfy	Satisfy	Neutral	Dissatisfy	Strongly dissatisfy
SeisTutor explained the content correctly	23	25	3	8	1
SeisTutor made the course as interesting as possible	31	19	9	1	0
The tutoring resources were adequate	21	19	7	9	4
The presentation of course content stimulated my interest during learning session	32	24	2	1	1
The course content are relevant and well organized	29	25	3	2	1
SeisTutor supported me to understand the content, which found confusing?	27	26	6	1	0
The quiz at the end of each week corresponds to the lessons taught?	28	27	3	1	1
The question wise hints were helpful	27	26	6	0	1
Did the SeisTutor react decidedly to your necessities?	26	21	11	1	1
Was the learning provided sufficiently to take the quiz?	36	18	4	2	0
During ongoing tutoring, assessments are a fair test of my knowledge and learning preferences	32	21	5	2	0

of people who used SeisTutor agreed that it helped them improve their learning performance and outcomes.

8.4.4 Kirkpatrick Phase 4: Evaluation of Results

This phase quantified the overall results in terms of effective learning. A paired sample T-test is performed on the learner's performance parameters, i.e., pretest and post-test scores of participants involved in both the studies (experimental and control groups). The hypothesis for inferencing the results is described as follows.

Case 1: The experimental group was subjected to a paired-sampled-T-test.

Hypothesis-Case-1.0:

The experimental group individuals have similar pretest and post-test mean values (negligible performance improvement).

Hypothesis-Case-1.1:

The experimental group members had distinct pretest and post-test average score (effective performance improvement).

Case 2: A paired-sampled-T-test was run on the control group.

Hypothesis-Case-2.0:

The pretest and post-test mean scores of the participants in the control group are comparable (negligible performance improvement).

Hypothesis-Case-2.1:

The pretest and post-test mean scores of the subjects in the control group differ (effective performance improvement).

The calculated T value (T_{Stats}), for the experimental group is 11.410, $P < 0.01$ (see Table 8.21). When compared to the score on the pretest, the average score on the post-test was 2.21786 points higher. It has been determined that hypothesis 1.0 cannot be accepted because the computed T stats value is greater than T critical. According to Table 8.19, there is a significant gap between the results of the pretest and the post-test (Table 8.20).

For the control group, the calculated T value (T Stats) is 5.312, P0.01 (see Table 8.22). The average score on the post-test was 1.24719 points, which is higher than the score on the pretest. Here, T stats is bigger than T critical, according to the calculations. So, we cannot accept hypothesis 2.0. Tables 8.20 and 8.22 show that there is a big difference between the scores on the pretest and the post-test.

Table 8.19 Statistical results of paired sample T-test of experimental group

Comparison item	Learning mode			
	Mean	N	Std. deviation	Std. error mean
Post-test of experimental group participants	3.9375	28	0.39455	0.07456
Pretest of experimental group participants	1.7196	28	0.99740	0.18849

Table 8.20 Statistical results of paired sample T-test of control group

Comparison item	Learning mode			
	Mean	N	Std. deviation	Std. error mean
Post-test of control group participants	3.6525	32	0.58915	0.10415
Pretest of control group participants	2.4053	32	1.39565	0.24672

Table 8.21 Paired-sampled-T-test results of experimental group

	Mean difference	Std. deviation	Std. error mean	95% confidence interval of the difference		T stats	df	T critics
				Lower	Upper			
Pair 1: Post-test of experimental group—Pretest of experimental group	2.21786	1.02856	0.19438	1.81902	2.61669	11.410	27	2.0518

Table 8.22 Paired-sampled-T-test results of control group

	Mean	Std. deviation	Std. error mean	95% confidence interval of the difference		T stats (Calc)	df	T critics
				Lower	Upper			
Pair 1: Post-test of control group—pretest of control group	1.24719	1.32804	0.23477	0.76838	1.72600	5.312	31	2.03951

Both groups have ruled out the null hypothesis, which means that both groups offer good training. But the goal of this analysis is to figure out which group has the biggest effect on improving the learning gain as a whole. The goal is reached by comparing the T stats of the two groups. The experimental group's T stats is greater than the control group's T stats. So, the experimental group has a big difference between their post-test scores and their pretest scores, and they also train better than the control group.

The experimental group outperforms the control group, according to this analysis, since it offers custom-tailored designed curriculum, tracks the emotions of students as they learn, and calculates the total level of comprehension [11].

8.5 Comparative Analysis of Performance Between the Learner-Centric Tutoring System "SeisTutor" with Existing Online Tutoring System

A comparative analysis is performed between the SeisTutor with the three open-source online learning system (my-Moodle, course-builder, and teachable). The inferences drawn from the learner's feedback are described below [7–9]. Tables 8.23, 8.24, 8.25 and 8.26 indicate the analysis of responses to learner feedback questionnaire for my-Moodle, course-builder, teachable, and SeisTutor, respectively.

Figures 8.16, 8.17, and 8.18 demonstrate the comparative analysis of the tutoring systems (teachable, course-builder, and my-Moodle) with the SeisTutor, on the basis of the strongly satisfactory level on a Likert scale of 1–5 ranging from strongly dissatisfied to strongly satisfy.

Table 8.23 Analysis of responses on learner feedback questionnaire: my-moodle

Parameters	Strongly dissatisfied (%)	Neutral (%)	Strongly satisfied (%)
GUI-based	24	30	46
Learner-centric learning environment	45	17	38
Dynamic profiling	35	18	46
Learning content	50	36	14
Resolving query during the session	53	13	34
Navigation support	22	27	51
Learner feedback	33	14	53
Cumulative percentage (%)	37.42	22.22	40.36

Table 8.24 Analysis of responses on learner feedback questionnaire: course-builder

Parameters	Strongly dissatisfied (%)	Neutral (%)	Strongly satisfied (%)
GUI-based	23	31	46
Learner-centric learning environment	84	12	4
Dynamic profiling	62	11	27
Learning content	80	4	16
Resolving query during the session	63	9	28
Navigation support	39	11	50
Learner feedback	39	15	46
Cumulative percentage (%)	55.74	13.33	30.92

Table 8.25 Analysis of responses on learner feedback questionnaire: teachable

Parameters	Strongly dissatisfied (%)	Neutral (%)	Strongly satisfied (%)
GUI-based	1	43	56
Learner-centric learning environment	80	16	4
Dynamic profiling	68	9	23
Learning content	93	4	3
Resolving query during the session	56	6	38
Navigation support	26	14	60
Learner feedback	26	10	64
Cumulative percentage (%)	50.10	14.47	35.43

Table 8.26 Analysis of responses of learner feedback questionnaire: SeisTutor

Parameters	Strongly dissatisfied (%)	Neutral (%)	Strongly satisfied (%)
GUI-based	16	21	71
Learner-centric learning environment	12	6	82
Dynamic profiling	10	7	83
Learning content	17	5	78
Resolving query during the session	23	9	68
Navigation support	20	12	68
Learner feedback	14	21	65
Cumulative percentage (%)	16	12	74

GUI-Based

According to the findings of the research, it has been determined that the percentage of students who report being strongly dissatisfied with SeisTutor is only 16%, however, this number rises to 24% when using my-Moodle. When compared to teachable, SeisTutor's neutral level of learner only accounts for 21% of students, while teachable raises that number to 43%. Learners report a significantly higher degree of strong satisfaction with SeisTutor (71%), in contrast to their experience with my-Moodle (46%), where this number was significantly lower.

Learner-Centric Learning Environment

From the analysis, it has been concluded that the strongly dissatisfaction level of the learner with the SeisTutor is 12%, while with the course builder, this percentage increases to 84%. The neutral level of learner with SeisTutor is 6%, while with the my-Moodle, this percentage increases to 17%. The strongly satisfaction level

Comparitive analysis of Intelligent Learning System

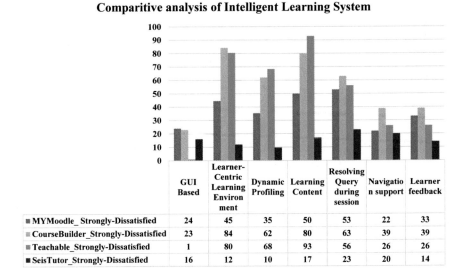

	GUI Based	Learner-Centric Learning Environment	Dynamic Profiling	Learning Content	Resolving Query during session	Navigation support	Learner feedback
■ MYMoodle_ Strongly-Dissatisfied	24	45	35	50	53	22	33
▨ CourseBuilder_Strongly-Dissatisfied	23	84	62	80	63	39	39
■ Teachable_Strongly-Dissatisfied	1	80	68	93	56	26	26
■ SeisTutor_Strongly-Dissatisfied	16	12	10	17	23	20	14

Fig. 8.16 Comparative studies of existing tutoring system with SeisTutor on strongly dissatisfaction level

Comparitive analysis of Intelligent Learning System

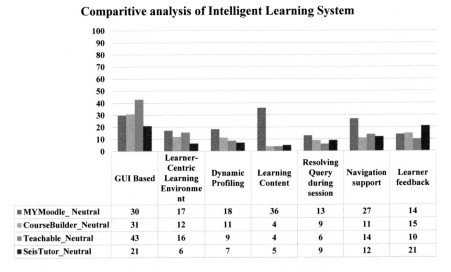

	GUI Based	Learner-Centric Learning Environment	Dynamic Profiling	Learning Content	Resolving Query during session	Navigation support	Learner feedback
■ MYMoodle_ Neutral	30	17	18	36	13	27	14
▨ CourseBuilder_Neutral	31	12	11	4	9	11	15
■ Teachable_Neutral	43	16	9	4	6	14	10
■ SeisTutor_Neutral	21	6	7	5	9	12	21

Fig. 8.17 Comparative studies of existing tutoring system with SeisTutor on neutral level

of learner with SeisTutor is 82%, while with the course builder and teachable, this percentage decreased to 4%.

Comparitive Analysis of Intelligent Learning System

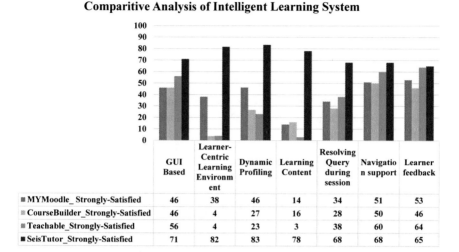

	GUI Based	Learner-Centric Learning Environment	Dynamic Profiling	Learning Content	Resolving Query during session	Navigation support	Learner feedback
■ MYMoodle_ Strongly-Satisfied	46	38	46	14	34	51	53
▪ CourseBuilder_Strongly-Satisfied	46	4	27	16	28	50	46
▪ Teachable_Strongly-Satisfied	56	4	23	3	38	60	64
■ SeisTutor_Strongly-Satisfied	71	82	83	78	68	68	65

Fig. 8.18 Comparative studies of existing tutoring system with SeisTutor on strongly satisfaction level

Dynamic Profiling

Based on the findings of the research, it has been determined that the percentage of learners who report being "strongly dissatisfied" with the SeisTutor is 10%, whereas the number of learners who report having this degree of dissatisfaction with teachable is 68%. When using my-Moodle, the learner's neutral level is significantly higher than when using SeisTutor, reaching 18% from a starting point of 7%. When using the teachable, the learner's level of strong satisfaction dropped from 83 to 23%, however, when using the SeisTutor, the learner's level of strong satisfaction was 83%.

Learning Content

According to the findings of the research, the percentage of students who are "strongly dissatisfied" with the SeisTutor product is only 17%, whereas the percentage of students who feel this way about the teachable product is 93%. When using Seis-Tutor, the learner's neutral level is just 5%, but when using my-Moodle, this number skyrockets to 36%. The percentage of learners who are strongly satisfied with the SeisTutor is 78%, whereas using teachable brings that percentage down to 3%.

Resolving Query During Session

According to the findings of the research, it has been determined that the level of learner dissatisfaction that is considered to be "strongly dissatisfied" with the Seis-Tutor is 23%, however, with the course builder, this number rises to 63%. When using my-Moodle, the percentage of students who have reached the neutral level is significantly higher than when using SeisTutor, which only has a nine percent success rate. When using the course builder, the learner's degree of strongly satisfied satisfaction dropped from 68 to 28%, whereas when using the SeisTutor, this percentage remained the same.

Navigation Support

Based on the findings of the research, it has been determined that the percentage of students who report being "strongly dissatisfied" with the SeisTutor is 20%, but the number of students who report having this degree of dissatisfaction with the course builder is 39%. When using my-Moodle, the percentage of students who have reached the neutral level is significantly higher than when using SeisTutor, which only has a neutral level achievement rate of 12%. While the learner's degree of strong satisfaction with the SeisTutor stands at 68%, this number dropped to 50% when they used the course builder.

Learner Feedback

Based on the findings of the research, it has been determined that the percentage of students who report being "strongly dissatisfied" with the SeisTutor is 14%, however, when using the course builder, this rate rises to 39%. When using SeisTutor, the learner has a neutral level of 21%; nevertheless, when using teachable, this proportion drops to 10%. When using the course builder, the learner's level of strongly satisfied satisfaction dropped from 65 to 46%, whereas when using the SeisTutor, this number remained at 65%.

According to the findings of this investigation, we are able to state that none of the tutoring systems that we have discussed up until this point had the capabilities of flexibility, dynamic profiling, and personalization. The most crucial aspects of SeisTutor are the program's capacity for personalization, adaptability, and dynamic profiling. It was discovered that 71% of learners are very happy with the GUI-based feature, 82% are very happy with the learning centric learning environment feature, 83% are very happy with the dynamic profiling feature, 78% are very happy with the learning content feature, 68% are very happy with the resolving query during session feature, 68% are very happy with the navigation support feature, and 65% are very happy with the learner feedback feature. The overall conclusion drawn from the comparative study is that the level of satisfaction of students using the SeisTutor is 74%, while their level of satisfaction using my-Moodle is 40.36%, while their level of satisfaction using teachable is 35.43%, and their level of satisfaction using course builder is 30.92%.

8.6 Conclusion and Future Scope

8.6.1 Summary

The objective of the presented work is to develop an adaptive tutoring engine, facilitating, knowledge base delivery through a learner-centric learning path. The design and development of an adaptive domain model and pedagogy model make the tutoring engine, an adaptive tutoring engine and provide the learner-centric learning path to the learner. The domain knowledge incorporated in ITS is seismic data interpretation, which is an experiential knowledge domain. Thus, acquiring, characterizing, sequencing, validating, and developing personalized course content (based on learner's learning profile and learning style), of the SDI knowledge domain creates a pool of adaptive knowledge base or repository. The adaptive pedagogy model leads to the systems that provide custom-tailored learning material to the learner based on the learner's prior knowledge, learning profile, and learning style. The custom-tailored learning path recommendation at the beginning of the learning session is rarely recommended in the existing ITS, this is due to the lack of empathy in ITS. Thus, this research aims to focus on the domain model and the pedagogy model. Therefore, the answer of the research questions is discussed below. The research questions drawn from the literature are–What are the steps involved to gather experiential knowledge from domain experts? How to represent experiential knowledge? On what criteria, learning material is aligned as per learner preference? How to generate a course coverage plan, which is exclusively designed for the learner? How a system can identify the learner preferences, exclusive course coverage plan, and give a custom-tailored pedagogical recommendation for adaptivity?

The following highlights the research contribution based on the conducted research work.

- An adaptive domain model indicates that the ITS offers an adaptive learning material that is offered as per the instruction received from the pedagogy model. Adaptive learning material specifies that the learning materials are aligned or structured as per the learner competency level and the learning preferences. As mentioned above, the seismic data interpretation domain is highly individualistic. Therefore, for gathering, causal maps and semi-structured acquisition techniques are utilized. After extraction of knowledge, knowledge manager sequences and classify the gathered knowledge as per seismologist's guidelines. The knowledge manager then validates the sequenced knowledge through ongoing consultation with seismologists and develops knowledge capsules. To make the adaptive domain model, the learning materials and restructured and realigned as per learning profile ("beginner," "intermediate," "expert") and learning style ("imagistic," "intuitive," "auditory," "active") adding up to twelve different combinations. Therefore, the tutoring engine offering the learner a customized learning experience by delivering tailored subject matter.
- An adaptive pedagogy model indicates that ITS offers a learner-centric learning path. The learner-centric learning path specifies that ITS offer personalized learning paths. This intelligent feature is implemented using the "BUG MODEL". The BUG MODEL is used to identify the learner's previous/prior knowledge by identifying the learner's bugs during the pretest. The presented novel approach in this book has the advantage to determine the learner prior knowledge and recommends the custom-tailored course coverage plan that improves the effectiveness of the system.

8.6.2 Conclusion

The presented research work focused on the development of a personalized, intelligent tutoring system for the domain "seismic data interpretation." This research work illustrates the design, development, and evaluation of the personalized intelligent tutoring system. The presented personalized intelligent tutoring system christened as SeisTutor, emulates the human cognitive intelligence by incorporating the artificial intelligence features, i.e., custom-tailored curriculum sequencing module, tutoring strategy recommendation module, CNN-based emotion recognition module, and performance analyzer module (degree of understanding module). Total of 60 learners have participated in the evaluation process. The participants were classified into two groups: control group and experimental group. Out of 60 participants, 32 of them designated as control group, and remaining 28 is designated as experimental group.

There are two aspects of the evaluation process, the first aspect is to identify which group prompts improvement in learning, and second aspects are to determine the learner's behavior, reaction, comfort level, and overall results. The first aspect of the evaluation process is accomplished by using the one-tailed ANOVA, and second aspect of the evaluation process is accomplished by using well accepted four phases/stage Kirkpatrick evaluation model.

ANOVA tests conducted on the pretest and post-test scores. The results indicate the effective learning gain of 44.34% by experimental group. The calculated F-ratio of the ANOVA test for the experimental group is 119.71, which is higher than the calculated F-ratio for control group 21.68. Thus, the experimental group is having a significant difference in the post-test and pretest scores and provides effective learning against the control group.

Kirkpatrick's four levels evaluation model is another widely accepted method for evaluating the effectiveness of the learning program. The levels are 1-reaction, 2-learning, 3-behavior, and 4-results. The outcome of reaction reveals that 44% of learners are happy with the offered learning content (customized learning content) and teaching process, i.e., pedagogy. The outcome of learning reveals that experimental group possesses 44.34% of learning gain and control group holds only 24.8%. Thus, the experimental group succeeds in enhancing the learner interest and curiosity, which indirectly increases the learner's performance. The outcome of behavior reveals that the presented system design produces productive learning in seismic data interpretation through incorporating computer science and artificial intelligence features. Besides, 86% of learners were satisfied and achieved better results with SeisTutor and improved their learning with the selection of appropriate adaptive tutoring strategies. The outcome of the results indicates that calculated T value (T_{Stats}), for the experimental group is 11.410, and control group is 5.312, $P <$ 0.01. The average post-test score of experimental group was 2.21786 points which are higher than pretest scores. The average post-test score of control group was 1.24719 points which are higher than pretest scores. Here, the calculated (T_{Stats}) is greater than $T_{critical}$, thus both the groups rejected hypothesis 1.0 and 2.0. Furthermore, the

T_{Stats} value of both the groups is compared. From the results, it has been revealed that the experimental group provides more effective learning against control group. In addition to this, SeisTutor is compared with the existing open-source tutoring system. From the analysis, it has been concluded that 74% of learner are strongly satisfied with the SeisTutor (*"GUI-based," "learner-centric learning environment," "dynamic profile," "learning content," "resolving learner query during session," "navigation support,"* and *"earner feedback"*).

8.6.3 Future Scope

The findings of the presented work in this book can be used for further research and development. In the accompanying sections, conceivable future headings are discussed.

Through the findings and discussion of the study's recommendation and future scope have been put forward, these are as follows:

- In future, the implementation of a domain-independent intelligent tutoring system will be a new sub-domain to be explored. The domain independence reduces the efforts of creating a whole ITS system.
- Extension of custom-tailored curriculum recommendation in the tutoring system will be the area to be worked on. There is a need to find out other intelligence techniques for determining learner's lack of knowledge of technical terms, which is discussed during the ongoing learning session. The custom-tailored curriculum module recommends the learner-centric learning path based on the prior knowledge of the learner at the beginning of the learning session.
- The presented research work in this book utilizes the CNN-based Emotion Recognition Module is used to track the learner's emotion during the ongoing learning session. In future, facial expression can be considered as a key parameter for pedagogy flipping (when the learner is not happy with the recommended pedagogy) during the ongoing learning session.
- The presented research work in this book considers only twelve tutoring strategy (the combination of one learning style with the one learning profile). In future, the combination of more than one learning style with one learning profile can be considered for ITS.

References

1. Singh, N., Kumar, A., Ahuja, N.J.: Implementation and evaluation of personalized intelligent tutoring system. Int. J. Innov. Technol. Explor. Eng. (IJITEE) **8**, 46–55 (2019)
2. Singh, N., Ahuja, N.J.: Bug model based intelligent recommender system with exclusive curriculum sequencing for learner-centric tutoring. Int. J. Web-Based Learn. Teach. Technol. (IJWLTT) **14**(4), 1–25 (2019)
3. Singh, N., Ahuja, N.J.: Implementation and evaluation of intelligence incorporated tutoring system. Int. J. Innov. Technol. Explor. Eng. (IJITEE) 8(10), 4548–4558
4. Singh, N., Ahuja, N.J.: Empirical analysis of explicating the tacit knowledge background, challenges and experimental findings. Int. J. Innovative Technol. Exploring Eng. (IJITEE) **8**(10), 4559–4568 (2019)
5. Ahuja, N.J., Singh, N., Kumar, A.: Development of knowledge capsules for custom-tailored dissemination of knowledge of seismic data interpretation. In: Networking Communication and Data Knowledge Engineering, pp. 189–196. Springer, Singapore (2018)
6. Kumar, A., Singh, N., Ahuja, N.J.: Learning styles based adaptive intelligent tutoring systems: document analysis of articles published between 2001 and 2016. Int. J. Cogn Res. Sci. Eng. Educ. **5**(2), 83 (2017)
7. Singh, N., Gunjan, V.K., Kadiyala, R., Xin, Q., Gadekallu, T.R.: Performance Evaluation of SeisTutor using cognitive intelligence-based "Kirkpatrick Model". Comput. Intell. Neurosci. (2022)
8. Singh, N., Gunjan, V.K., Nasralla, M.M.: A parametrized comparative analysis of performance between proposed adaptive and personalized tutoring system "seis tutor" with existing online tutoring system. IEEE Access **10**, 39376–39386 (2022)
9. Singh, N., Gunjan, V.K., Mishra, A.K., Mishra, R.K., Nawaz, N.: SeisTutor: a custom-tailored intelligent tutoring system and sustainable education. Sustainability **14**(7), 4167 (2022)
10. Ahuja, N.J., Singh, N., Kumar, A.: Adaptation to emotion cognition ability of learner for learner-centric tutoring incorporating pedagogy recommendation. Int. J. Control Theory Appl. **9**(44), 15–30 (2016)
11. Singh, N., Ahuja, N.J., Kumar, A.: A novel architecture for learner-centric curriculum sequencing in adaptive intelligent tutoring system. J. Cases Inf. Technol. (JCIT) **20**(3), 1–20 (2018)

Appendix A

Keyword for Week 1

Main reference matric	N-gram co-occurrence matric	Labels
Petroleum exploration	Four	Basic stages
	Seismic acquisition, seismic data processing, Understanding the data and Seismic data interpretation	Stage classification
Seismic data processing	Various seismic gathers	Types of analysis
	Common midpoint binning	Types of analysis
	Stacking	Types of analysis
Seismic waves	Sudden breaking of rocks	Reason of occurrence
	Two types	Classification
Body waves	Primary waves	Classification types
	Secondary Waves	Classification types
Surface waves	Rayleigh waves	Classification types
	Love waves	Classification types
Primary waves	Compressional waves	Wave motion
	Fastest waves	Speed
	Solid and liquid	Can move
Secondary waves	Transverse wave	Wave motion
	Medium	Speed
	Solid	Can move
Rayleigh waves	Rolls	Wave motion
	Slow	Speeds
	Earth surface	Can travels
Love waves	Side to side	Wave motion

(continued)

N. Singh et al., *Cognitive Tutor*, Advanced Technologies and Societal Change, https://doi.org/10.1007/978-981-19-5197-8

(continued)

Main reference matric	N-gram co-occurrence matric	Labels
	Fastest	Speeds
	Earth surface	Can travels
Velocity	Reflection	Change result
	Refraction	Change result

Keyword for Week 2

Main reference matric	N-gram co-cccurrence matric	Labels
Seismic sources	Marine	Acquisition
	Land	Acquisition
Land acquisition	Explosives	Technique used
	Thumper truck	Technique used
	Vibroseis	Technique used
Marine acquisition	Airgun	Technique used
Explosives	Drill Hole	Perform by
	Chemical composition	Made up of
	Cheap	Cost
Thumper truck	Vibrator	Process
	seismic waves	Emit
Vibroseis	Vibrator	Process
	seismic waves	Emit
Airgun	pneumatic chambers	Consist of
	Air	Release
Seismic receivers	Marine	Acquisition
	Land	Acquisition
Land acquisition	Geophone	Technique used
Marine acquisition	Hydrophone	Technique used
Geophone	Electric voltage	Produce
	Coil and magnet	Relative motion
Hydrophone	Change	Detect
	Pressure sensor	Compasses tail buoys
Analog recording	Analog amplifier	Analog to digital converter process
	Analog filter	Analog to digital converter process
	Magnetic analog recorder	Analog to digital converter process
	Continuous	Signals form
Analog amplifier	Automatic gain control	Special circuit
Analog filter	Noise	Remove

(continued)

(continued)

Main reference matric	N-gram co-cccurrence matric	Labels
Magnetic analog recorder	Modulators	Seismic range
Horizon	Contouring	Steps involve
	Well-Calibration	Steps involve
	Velocity estimation	Steps involve
	Depth map	Steps involve
Contouring	Three dimensional form	Representation of horizon
Well-Calibration	Two way travel time to depth	Conversion
Velocity estimation	Migration	Seismic Imaging
Depth map	Structural ambiguity	Used to remove

Keyword for Week 3

Main reference matric	N-gram co-occurrence Matric	Labels
Faults	Foot wall	Walls
	Hanging wall	Walls
	Normal faults	Types
	Reverse faults	Types
	Strike slip faults	Types
	Stress	Reason of breaking
Normal faults	Divergent	Plate motion
	Tension	force
Reverse faults	Convergent	Plate motion
	Pushing	force
Strike slip faults	Transform	Plate motion
Stress	Tensile	Types
	Compressive	Types
	Shear	Types
Strain	Tensile	Types
	Compressive	Types
	Shear	Types
	Volumetric	Types

Keyword for Week 4

Main reference matric	N-gram co-occurrence matric	Labels
Direct hydrocarbon indicators	Flat spot	Types
	Bright spot	Types
	Dim spot	Types
Velocity interpretation	Average	Types
	Interval	Types
	Root mean square	Types
Faults map	Understanding	Uses
	Drilling plan	Uses
	Location	Uses
2D	Less data	Survey
3D	Data	Survey

Appendix B

See Tables B.1, B.2, B.3, B.4.

Table B.1 Learners' feedbacks on learning through my-moodle

	Questions	Strongly-dissatisfied	Neutral	Strongly-satisfied
1	How satisfied are you with the look and feel (user interface design) of this system?	17	21	32
2	The pre tutoring test is conducted	25	13	32
3	As a learner, did you feel that your learning style was appropriately judged?	32	20	18
4	The information provided by my-moodle is at a level that you understand	31	9	30
5	Were you convenient and satisfied with the tutoring strategy presented to you by my-moodle?	25	10	35
6	Based on your prior subject knowledge, has my-moodle accurately determined exclusive curriculum for you?	33	12	25
7	The course content are relevant and well organized	35	25	10
8	Is my-moodle accurately determined your psychological (emotional) state during tutoring session? (give rating)	25	3	42
9	The understanding test at the end of each week corresponds to the lessons taught?	35	10	25

(continued)

N. Singh et al., *Cognitive Tutor*, Advanced Technologies and Societal Change, https://doi.org/10.1007/978-981-19-5197-8

Table B.1 (continued)

	Questions	Strongly-dissatisfied	Neutral	Strongly-satisfied
10	The post tutoring evaluation system (week wise understanding) as it exists is:	24	23	23
11	The system navigation support enabled you to find the needed information	15	19	36
12	Is my-moodle Handle the learner's Issue/ during the ongoing learning session	37	9	24
13	Is my-moodle gathers the learner valuable feedback	23	10	37

Table B.2 Learners' feedbacks on learning through course-builder

	Questions	Strongly-dissatisfied	Neutral	Strongly-satisfied
1	How satisfied are you with the look and feel (user interface design) of this system?	16	22	32
2	The pre tutoring test is conducted	65	5	0
3	As a learner, did you feel that your learning style was appropriately judged?	50	18	3
4	The information provided by course-builder is at a level that you understand	63	4	3
5	Were you convenient and satisfied with the tutoring strategy presented to you by course-builder ?	65	4	1
6	Based on your prior subject knowledge, has course-builder accurately determined exclusive curriculum for you?	56	9	5
7	The course content are relevant and well organized	56	3	11
8	Is course-builder accurately determined your psychological (emotional) state during ongoing tutoring session? (give rating)	61	7	2
9	The understanding test at the end of each week corresponds to the lessons taught?	61	7	2
10	The post tutoring evaluation system (week wise understanding) as it exists is:	4	12	54

(continued)

Table B.2 (continued)

	Questions	Strongly-dissatisfied	Neutral	Strongly-satisfied
11	The system navigation support enabled to find the needed information	27	8	35
12	Is course-builder handle the learner Issue/ problem during the ongoing learning session	44	6	20
13	Is course-builder gathers the learner valuable feedback	27	11	32

Table B.3 Learners' feedbacks on learning through teachable

	Questions	Strongly-dissatisfied	Neutral	Strongly-satisfied
1	How satisfied are you with the look and feel (user interface design) of this system?	1	30	39
2	The pre tutoring test is conducted	66	4	0
3	As a learner, did you feel that your learning style was appropriately judged?	47	18	5
4	The information provided by teachable is at a level that you understand	67	1	2
5	Were you convenient and satisfied with the tutoring strategy presented to you by teachable ?	65	3	2
6	Based on your prior subject knowledge, has teachable accurately determined exclusive curriculum for you?	63	5	2
7	The course content are relevant and well organized	65	3	2
8	Is teachable accurately determined your psychological (emotional) state during tutoring session? (give rating)	66	1	4
9	The understanding test at the end of each week corresponds to the lessons taught?	39	28	3
10	The post tutoring Evaluation system (Week Wise Understanding) as it exists is:	12	13	45
11	The system navigation support enabled to find the needed information	18	10	42

(continued)

Table B.3 (continued)

	Questions	Strongly-dissatisfied	Neutral	Strongly-satisfied
12	Is teachable handle the learner Issue/ problem during the ongoing learning session	39	4	27
13	Is teachable gathers the learner valuable feedback	18	7	45

Table B.4 Learners' feedbacks on learning through SeisTutor

Questions	Strongly-dissatisfied	Neutral	Strongly-satisfied
How satisfied are you with the look and feel (user interface design) of this system?	11	10	50
The pre tutoring test is conducted	6	5	59
As a learner, did you feel that your learning style was appropriately judged?	11	5	55
The information provided by SeisTutor is at a level that you understand	13	6	52
Were you convenient and satisfied with the tutoring strategy presented to you by SeisTutor?	8	3	60
Based on your prior subject knowledge, has SeisTutor accurately determined exclusive curriculum for you?	8	4	58
The course content are relevant and well organized	12	4	55
Is SeisTutor accurately determined your psychological (emotional) state during tutoring session? (give rating)	8	7	55
The understanding test at the end of each week corresponds to the lessons taught?	4	5	62
The post tutoring evaluation system (week wise understanding) as it exists is:	6	3	61
The system navigation support enabled to find the needed information	14	8	48
Is SeisTutor handle the learner Issue during the ongoing learning session	16	6	48
Is SeisTutor gathers the learner valuable feedback	10	15	46

Printed in the United States
by Baker & Taylor Publisher Services